Industry 4.0 Topical Conference 2024

Topical Conference at the 2024 AIChE Spring Meeting and 20th Global Congress on Process Safety

New Orleans, Louisiana, USA
24-28 March 2024

ISBN: 979-8-3313-1712-6

Printed from e-media with permission by:

Curran Associates, Inc.
57 Morehouse Lane
Red Hook, NY 12571

AIChE
120 Wall Street, FL 23
New York, NY 10005-4020

Phone: (800) 242-4363
Fax: (203) 775-5177

www.aiche.org

Additional copies of this publication are available from:

Curran Associates, Inc.
57 Morehouse Lane
Red Hook, NY 12571 USA
Phone: 845-758-0400
Fax: 845-758-2633
Email: curran@proceedings.com
Web: www.proceedings.com

TABLE OF CONTENTS

POSTER SESSION: INDUSTRY 4.0/ANALYTICS AND AI

ROBOTICS, CYBERPHYSICAL, AND AUTONOMOUS SYSTEMS

INDUSTRY 4.0 - SENSORS AND ANALYZERS FOR ONLINE MEASUREMENTS

PANEL DISCUSSION: NLP AND LLM IN THE CHEMICAL INDUSTRY

VENDORS' PERSPECTIVE (SESSION SPEAKERS INVITED)

128a AI for Predictive Maintenance with or without Vibration Sensing

Iiro Esko, David Peterson

New technologies such as AI must fit in, and add value, into legacy plant operations. During the last years many obstacles and objections have been overcome, such as:

- Putting in more sensors can be a showstopper.
- Plenty of asset health and AI/ML software demand vibration sensor data.
- Artificial Intelligence and Machine Learning applications can give insights to their users even without vibration data.
- Investment justification is possible when a logical argument can be made around the right data being available to reach an investment decision based on ROI.

This abstract covers angles regarding process industries asset health analytics and AI applications with minimal vibration data availability. It is backed by multiple installations in US facilities across different industries and focuses on the bottom-line impact of early detection of anomalies of large fleets of assets through application of priority attention index to bring items to the attention of front-line maintenance workers.

By joining this session, you will learn about trends in different industries, enabling returns from AI predictive maintenance systems, deployment across a wide range of assets and asset types, as well as easy utilization by current organization and employees of your organization. Example architecture of an AI system for predictive maintenance is evaluated from data sources to a core platform, to operational contextualization, to anomaly detection, and finally to prioritized notifications for maintenance-responsible colleagues.

Practical considerations regarding rotating equipment, fixed assets, device and sensor data for process and equipment are included, as are scaling at a specific site location and globally - with a focus on impact when it comes to sustainable operations, mobile working, related risks and costs, increased availability, and precise execution of predictive maintenance. Another practical focus of this abstract is what is possible with limited or absent vibration data.

Most maintenance practices are based on service-interval schedules, a preventative and reactive approach that doesn't take actual machine or equipment usage and health into account, and which is questionable in substantially reducing unplanned downtime and improving maintenance effectiveness. Predictive maintenance (PdM), on the other hand, is a proactive approach, enabling machine and equipment failures to be dealt with before they stop production. PdM is achieved by analyzing huge volumes of machine and maintenance data to decode the health of machines and equipment and enable maintenance staff to optimize their activities. By predicting break downs, companies can therefore eliminate sudden failures, reduce unplanned downtime, increase machine life, optimize scheduled maintenance, and reduce the routine replacement of parts that may, in fact, be perfectly healthy.

This abstract is aimed at intermediate level participants when it comes to AI and predictive maintenance ranging from front-line workers to experts to managers and directors.

128b Evaluation of the Smart Process Analysis Framework in Industrial Data

Dustin Kenefake, Swee-Teng Chin, Wenzhao Yang, Ivan Castillo, María Paz Ochoa

Recently, there has been a marked increase in the use of automated modeling tools for wide ranging applications, such as time series forecasting, soft sensors, and predictive modeling. One of these tools, Smart Process Analysis (SPA), has been proposed to be a solution for automated modeling of manufacturing and process data. This tool aims to suggest model types and model regression, with cross validation leading the user thru the process. The proposed tool is benchmarked on some synthetic test cases, as well as an industrial case study on the effect of an additive to the high temperature performance asphalt. The results and discussions of the work will be presented in this talk.

128c Integrating Process Safety, Risk Assessment, AI, and Systems Biology to Regulate Human Health and Eco-Risks Arising from Industrial Accidents and Occupational Exposure

Achilleas Karakoltzidis, Nafsika Papaioannou, Ilias Frydas, Thanasis Papageorgiou, Dayna Schultz, Catherine Gabriel, Elisavet Renieri, Spyros Karakitsios, Dimosthenis Sarigiannis

In this study, a holistic approach is introduced for assessing the potential impact of industrial accidents or daily occupational exposure on worker well-being with the associated short- and long-term health implications. Our methodology leverages mechanistic models and machine learning techniques from numerous research fields including process safety, computational fluid dynamics, toxicokinetic models, multimedia models and artificial intelligence for data gap filling. To illustrate the application of our approach, a use case is developed herein, focusing on bisphenol A, a chemical of widespread use and of significant concern due to its characterization as an endocrine disruptor and its association with a plethora of metabolic syndromes and adverse outcomes. Bisphenol A finds extensive use in food packaging, thermal paper, and as an essential component in plastics manufacturing, with a global annual production estimated at approximately 7,000 kt.

Bisphenol A is primarily synthesized through the condensation reaction of phenol and acetone with the presence of an acid catalyst. Despite its apparent simplicity, this production process carries a fire hazard due to the highly flammable nature of the involved materials. The scope of accidents investigated in this industry pertains to fire incidents, encompassing all stages of the production process. Given its relatively high molecular weight, bisphenol A is non-volatile and only investigated for pool fire accidents, in contrast to the raw materials explored for flash fires, jet fires, and explosions. To assess the industrial-level consequences of potential accidents, we employed the DNV software package, Phast, which facilitated both quantitative analysis and computational fluid dynamics modeling. This approach allowed us to calculate and visualize the dispersion of hazardous plumes across the entire facility. Furthermore, utilizing spatial integration and fluid dynamics insights provided by the software, we were able to position hypothetical workers along the length of the installation, enabling the consideration of multiple exposure scenarios for quantitative risk assessment.

The functionalities of the Phast software are valuable for calculating toxicological consequences, offering insights into the probability of life-threatening outcomes using specific coefficients with the Probit function. However, for a more comprehensive examination for the impact of an accident on the health of individuals exposed, a deeper investigation is necessary. In our quest to advance this analysis, we have incorporated the INTEGRA software package into the developing methodological pipeline (Sarigiannis et al., 2014; Sarigiannis et al., 2016). INTEGRA empowers users to explore the entire trajectory from the initial exposure event to the environmental context (source-to-dose-continuum). With a detailed level three multimedia model, INTEGRA facilitates an in-depth environmental assessment of the potential consequences of an accident. This environmental assessment proves particularly vital as it evaluates the impact on micro-environments within the affected area. Leveraging the CFD model provided by Phast, we can precisely determine the concentration of bisphenol A (BPA) across the three primary exposure routes: inhalation, oral ingestion, and skin contact. This information, in turn, allows us to employ the detailed Physiologically Based Toxicokinetic (PBTK) model provided by the platform. Through the PBTK model, we convert external doses in the three exposure routes into internal doses, a crucial step that enables us to delve into molecular and organ-level interactions between xenobiotics and endogenous factors.

To achieve this, we integrate data from both in vivo and in vitro studies, along with human biomonitoring data. This comprehensive dataset is then enriched with metabolomics and transcriptomics analyses specific to bisphenol A (BPA), including numerous concentration levels corresponding to different exposure scenarios. From these data, we construct integrated dose-response curves, effectively relating concentration levels to interactions with a range of endogenous biomarkers. The next step involves the development of a comprehensive systems biology model capable of encompassing a wide array of biological mechanisms to quantify the potential impact of BPA exposure. To achieve this, multi-omics data are combined to facilitate a joint pathway analysis. These biological pathways are then integrated into the methodology presented by Karakoltzidis et al. (2023) to introduce Next Generation Systems Biology models. These models consist of ordinary differential equations accounting for all endogenous factors within a biological system, including proteins, enzymes, and endogenous metabolites. This approach offers a holistic understanding of the intricate interactions and responses within the biological system resulting from BPA exposure.

Parameterizing such a complex mathematical model presents significant challenges. To address this, we harnessed data from a multitude of publicly accessible databases, including BRENDA, SABIO RK, HMDB, BioModels, UniProt, KEGG, and more. We developed in-house tools to compile and organize this wealth of information, resulting in expansive datasets, totaling many gigabytes. These datasets served as the foundation for the development of AI models. Using Deep Learning architectures, we constructed three models capable of accurately predicting kinetic parameters, including Michaelis-Menten constants and turnover numbers. All these models achieved impressive performance, with R-squared values exceeding 0.65, signifying their robust predictive capabilities. For the calculation of initial concentrations, a Machine Learning methodology was employed. It involved the use of unified networks to establish regression correlations, estimating the potential concentrations of unknown metabolites based on their interactions with reactants and products. In cases where training temporary unsupervised models led to overfitting or underfitting, data augmentation techniques were applied to enrich the dataset, ensuring the reliability and comprehensiveness of the model's predictions.

To establish a quantitative relation between the model including biomarkers and phenotypic diseases, we applied Natural Language Processing (NLP) techniques. We developed three token classification models, one for diseases, one for endogenous metabolites, and one for biomedical text mining, which could identify a wide array of biological entities, including DNA, proteins, cell types, cell lines, genes, and RNA. These models demonstrated excellent performance, with F1 scores surpassing 0.7; they were trained based on publicly available data.

Before collecting the model outputs, the results underwent in-house evaluation. Using these NLP tools, we scoured over 28,000 articles in PubMed, ultimately identifying 143 studies that linked endogenous metabolites to disease phenotypes. To further refine our focus, we concentrated on metabolic diseases such as obesity and different types of diabetes. We did that because BPA is deemed by the toxicological community to be highly linked with these outcomes. Consequently, by conducting a series of simulations, we could not only pinpoint BPA levels in the human body but also assess whether an accident could trigger adverse outcomes for the workers. This holistic approach provides a comprehensive understanding of the potential health impacts of BPA exposure enabling an on-site assessment that encompasses both short-term and long-term consequences of accidents, providing a comprehensive view of potential risks. Furthermore, industries can use this methodology to establish regular occupational health assessments, thus fostering a healthier and more efficient workforce. This dual benefit, combining safety and worker well-being, offers a comprehensive solution for industries seeking to improve their operations and safeguard their employees.

Karakoltzidis, A., Karakitsios, S., & Sarigiannis, D. (2023). MINER: A Text Mining Approach for the

Introduction of Next-Generation Systems Biology Models. *Toxicology letters, 384*. https://doi.org/10.1016/S0378-4274(23)00538-6

Sarigiannis, D., Karakitsios, S., Gotti, A., Loizou, G., Cherrie, J., Smolders, R., De Brouwere, K., Galea, K., Jones, K., Handakas, E., Papadaki, K., & Sleeuwenhoek, A. (2014). Integra: From global scale contamination to tissue dose. Proceedings - 7th International Congress on Environmental Modelling and Software: Bold Visions for Environmental Modeling, iEMSs 2014,

Sarigiannis, D. A., Karakitsios, S. P., Handakas, E., Simou, K., Solomou, E., & Gotti, A. (2016). Integrated exposure and risk characterization of bisphenol-A in Europe. *Food and Chemical Toxicology*, *98*, 134-147. https://doi.org/10.1016/j.fct.2016.10.017

23a The Challenges of Operationalizing Machine Learning/AI in Industry

Brent Railey

The exploding field of artificial intelligence (AI) and machine learning (ML) promises transformative potential for industry, heralding a potential new era of efficiency and innovation. However, the journey from theoretical models to practical applications is fraught with challenges that often temper the initial excitement. Even with successful Proofs-of-Concept in practical, real-world opportunities, operationalization and long-term maintenance can make emotions from "tempered" to "frustrated." This talk will discuss the realities concerning the challenges of bringing AI research and experimentation to operational applications across various types of AI: inferential, optimization, and Generative AI.

42b Unleashing the Power of Hybrid Modeling for Process Design and Optimization.

Reza Haghpanah, Brad Metzler, Evan Bergman, Aayush Singh, David Vickery

Process optimization via mathematical modeling is a powerful tool in the chemical process industry for developing and implementing process improvements. For example, mathematical models can be used to simulate different process scenarios and identify the optimum operating conditions that will maximize productivity while minimizing energy usage. This can lead to significant energy and cost savings, increased capacity, and profitability for the industry. However, developing accurate first principles (physics-based) models can be complex, especially when modeling large and complex processes with multiple physical and chemical phenomena. Hybrid models, which combine physics-based models with machine-learning models or data-driven models, can be used to address some of these limitations. By combining different modeling techniques, hybrid models can provide a more accurate and comprehensive representation of the process while reducing the computational cost of the model. This work presents the use of Aspen Hybrid Models^{TM} at Dow in a highly integrated process flowsheet. First-principles-driven hybrid models were developed in Aspen Plus V12.1 for reactors with complex reaction networks coupled with distillation columns and multiple recycle streams. Methodology for development of the hybrid model will be shared along with high level results of model validation versus plant performance and model optimization opportunities.

42c Harmonization of Process and Industrial Safety Principles in the Development of Green Chemicals and Materials: Pioneering the Safe and Sustainable By Design Paradigm

Achilleas Karakoltzidis, Fotini Nikiforou, Spyros Karakitsios, Dimosthenis Sarigiannis

In 2019, the European Commission introduced the EU Green Deal, an ambitious initiative aimed at transforming the European Union into a sustainable and climate-neutral continent. This comprehensive policy framework encompasses a wide array of strategies and measures to address climate change, safeguard the environment, and foster economic growth. The EU Green Deal outlines specific targets, including achieving carbon neutrality by 2050, enhancing energy efficiency, and promoting renewable energy sources. It also places a strong emphasis on conserving biodiversity, promoting sustainable agriculture, and forwarding a circular economy. Within the context of the EU Green Deal and the European Chemicals Strategy for Sustainability (European Commission, 2021), the concept of Safe and Sustainable by Design (SSbD) was introduced.

SSbD encompasses the entire life cycle of chemicals and materials, addressing safety and sustainability considerations from manufacturing to distribution, consumption and disposal. This approach encompasses aspects such as raw materials, safety procedures, recycling, use, and disposal. The application of SSbD principles to chemicals and materials is a significant step toward advancing environmentally friendly industrial procedures, with the European Union (EU) serving as a global benchmark for safety and sustainability objectives. In alignment with these objectives, guidance will be offered for establishing criteria that ensure compliance with these principles as well as fostering innovation to reduce or replace hazardous substances in accordance with emerging regulatory requirements.

As part of the EU Partnership for the Assessment of Risk of Chemicals (PARC) (Marx-Stoelting et al., 2023), a central aim is to establish a concrete and FAIRified (Findable, Accessible, Interoperable, Reusable) platform for assessing the safety of chemicals or materials concerning human health and their environmental sustainability during the development phase. This ambitious objective entails the implementation of data integration techniques and the incorporation of a variety of tools into a standardized methodological framework. These tools encompass a broad spectrum of applications, including Green, Circular, and Sustainable Chemistry, Green Process Design and Engineering, Life Cycle Impact Assessment, Risk and Toxicity Assessment, as well as the alignment with the EU Industrial Strategy, among others.

Process safety, which is frequently neglected during the initial material design phase, represents a comprehensive approach that involves recognizing, evaluating, and addressing hazards throughout a product or process's entire life cycle. A decisive component of process safety is the quantitative risk assessment, aimed at minimizing the chances of unforeseen accidents or failures at industrial settings. This proactive approach not only curtails the need for costly rework but also conserves resources in the long term. Furthermore, it promotes a culture of ongoing enhancement in materials design, emphasizing both industrial safety and environmental sustainability. The PARC SSbD toolbox incorporates a range of functions and hazard estimation tools for precise incident calculations. This includes the assessment of fireball incidents (Martinsen & Marx, 1999), jet and pool fire accidents (Cook et al., 1990), and the estimation of thermal radiation curves following the TNO Yellow Book principles (TNO Yellow Book, 2005). Explosion scenarios are implemented by applying TNO Yellow Book principles (TNO Yellow Book, 2005) as well.

It is evident that the routines described do not provide a comprehensive depiction of potential accidents or the severity of incidents resulting from process failures. To address this limitation, the SSbD toolbox incorporates a process-safety-smart-decision system. During the conceptual design of new ideas or the evaluation of alternative products, this system estimates a safety risk index. This schema relies on chemical structural information, conditional material properties, and calculations from the models mentioned earlier. Strictly speaking, EU PARC SSbD toolbox process safety functionality accepts a limited number of inputs, including flash and boiling point normalized by a reference temperature, flammability, explosiveness (determined using the previously described model outputs normalized with the total mass, lower flammability, and explosive limits, and the NFPA 704 fire diamond principle), as well as process conditions (temperature, pressure, type of inventory), and the rest of the parameters required to run the corresponding models. The data combination process yields normalized and dimensionless factors, which are subsequently added together to compute the corresponding safety index. The higher the index is, the more unsafe the industrial installation. The main objective here is not to provide a final assessment of industrial safety but to enable a comparison basis of numerous alternatives with the initial product for possible substitution or improvement. A more extensive assessment of safety aspects is reserved for a later stage.

A prominent point of discussion revolves around the high number of parameters required to operate the SSbD-smart-decision system, which becomes particularly pertinent for chemicals or materials that are relatively new or/and have not undergone prior evaluations in this context. To tackle this challenge, advanced 3D Quantitative Structure-Activity Relationship (QSAR) methodologies were employed, involving the introduction of artificial intelligence (AI) models for each property of interest. DIPPR's Project 801 Database is incorporated in the property estimation procedure. This approach guarantees that the decision system can apply the relevant methodology even in the absence of data during the development phase. It is noteworthy that the AI models developed thus far have demonstrated R-squared values exceeding 0.8.

Toxicity is assessed with advanced 3D deep learning QSARs, and Probit functions that are parametrized with QSARs as well, if it is needed. Furthermore, concerning the human health safety and the quantitative point of view, a thorough examination of the source to internal dose continuum is facilitated by the incorporation of the INTEGRA platform (Sarigiannis et al., 2014; Sarigiannis et al., 2016) into the SSbD toolbox. This approach employs level three multimedia models, detailed exposure, and occupational exposure models, as well as comprehensive Physiology-based Pharmacokinetic (PBPK) models to evaluate associated risks. Lastly, the assessment of product and process life cycles in the first version of the software is conducted by using the brightway2 framework, a publicly accessible software (Mutel, 2017).

Ensuring the principles of Safe and Sustainable by Design (SSbD) for innovative materials extends beyond only considering chemical structures and their life cycle assessment. It involves a comprehensive examination of the processes, including possible alternative methods, from raw material application to the final product. This approach allows users and industries to gauge the impact of potential accidents and alternative processes. As we design and conceptualize SSbD chemicals and alternatives, we also evaluate the quality and risks associated with the production processes. The presented methodology minimizes the chances of severe accidents, reduces energy consumption, lessens environmental burdens, and enhances production efficiency by optimizing the chemical structures. Our focus is not only on reducing the possibility of leaks through advanced monitoring systems but also on designing green processes to ensure that by-products meet environmental standards and are safe for the human health. This approach also improves working conditions over the long term, enhancing overall well-being. In summary, process safety is not just a regulatory requirement; it is a foundational element for achieving safe and sustainable by design materials. By integrating process safety into materials innovation, we can mitigate risks for the human health, promote environmental sustainability, and encourage innovation.

Cook, J., Bahrami, Z., & Whitehouse, R. (1990). A comprehensive program for calculation of flame radiation levels. *Journal of loss prevention in the process industries*, *3*(1), 150-155. https://doi.org/10.1016/0950-4230(90)85039-C

European Commission. (2021). *COMMUNICATION FROM THE COMMISSION TO THE EUROPEAN PARLIAMENT, THE COUNCIL, THE EUROPEAN ECONOMIC AND SOCIAL COMMITTEE AND THE COMMITTEE OF THE REGIONS Pathway to a Healthy Planet for All EU Action Plan: 'Towards Zero Pollution for Air, Water and Soil'*. Retrieved from https://eur-lex.europa.eu/legal-content/EN/TXT/?uri=CELEX%3A52021DC0400&qid=1623311742827

Martinsen, W. E., & Marx, J. D. (1999). An improved model for the prediction of radiant heat from fireballs. 1999 International Conference and Workshop on Modeling Consequences of Accidental Releases of Hazardous Materials San Francisco, California September,

Marx-Stoelting, P., Rivière, G., Luijten, M., Aiello-Holden, K., Bandow, N., Baken, K., Cañas, A., Castano, A., Denys, S., & Fillol, C. (2023). A walk in the PARC: developing and implementing 21st century chemical risk assessment in Europe. *Archives of Toxicology*, 1-16. https://doi.org/10.1007/s00204-022-03435-7

Mutel, C. (2017). Brightway: an open source framework for life cycle assessment. *Journal of open source software*, *2*(12), 236. https://doi.org/10.21105/joss.00236

Sarigiannis, D., Karakitsios, S., Gotti, A., Loizou, G., Cherrie, J., Smolders, R., De Brouwere, K., Galea, K., Jones, K., Handakas, E., Papadaki, K., & Sleeuwenhoek, A. (2014). Integra: From global scale contamination to tissue dose. Proceedings - 7th International Congress on Environmental Modelling and Software: Bold Visions for Environmental Modeling, iEMSs 2014,

Sarigiannis, D. A., Karakitsios, S. P., Handakas, E., Simou, K., Solomou, E., & Gotti, A. (2016). Integrated exposure and risk characterization of bisphenol-A in Europe. *Food and Chemical Toxicology*, *98*, 134-147. https://doi.org/10.1016/j.fct.2016.10.017

TNO Yellow Book. (2005). Methods for the Calculation of the Physical Effects of the Escape of Dangerous Material. In: TNO Rijswijk, The Netherlands.

68a Industry 4.0 and Remote Monitoring: Navigating the Challenges of Interconnectivity and Cybersecurity

Benoit Janvier, Christian Hager

In an era characterized by escalating cyber threats and data vulnerabilities, there is a vital need for manufacturers to safeguard the integrity and security of their control networks. However, this imperative often clashes with the growing demand for real-time interconnectivity and digitization, particularly as industries seek remote technical support from OEMs and technical experts.

This session will present different connectivity options available for remote cyber-secured support, exploring how "manufacturers" can maintain a low latency real-time operational data flow for monitoring, advanced analytics, flow modeling, energy optimization and predictive maintenance while taking advantage of robust cybersecurity measures, benefiting from the advantages of interconnectivity between the enterprise (IT) network and the operational technology(OT) systems.

Additionally, the speakers will present the capabilities of small form-factor data diodes as a solution to establish a strict, cyber physical, secure-by-design unidirectional data flow to enable a monitoring-only approach without introducing additional threat vectors.

Attendees will learn from multiple use/business cases how large chemical and manufacturing operations have protected their operational technology and increased real-time monitoring capabilities within the enterprise/IT network without exposing the critical infrastructure to external threat actors and internal vulnerabilities.

The session will also impart the power of expanded analytics with real-time data access to areas that were previously "air gapped" or part of the industrial control systems.

Further, several use cases will highlight the ability of taking operational data from secure facilities, equipment or networks and securely transporting that data to a cloud instance (e.g. AWS or on-prem) implementing alert management and energy efficiency processes to ultimately reduce the total cost of ownership and increase the return on investment of cyber physical security as part of a defense-in-depth strategy.

Comprehensive Compliance Management: A necessary tool for the effective management of Industrial Cybersecurity

Juan C. Angarita Castellanos, CEO (and leader of services for the Oil, Gas, and Chemical industry), IMS Global. Petroleum Engineer (UIS), MSc Computer Science (UIS); Certified Functional Safety engineer; Certified Lead auditor/assessor in several standards related to IT/OT/IoT and industrial Asset management, Information Security, IT/OT/IoT Cybersecurity, Cloud Security, Privacy, Business Continuity, IT/OT Services, Artificial Intelligence, Software Quality & Security, Quality, and Data Centers; Certified Lead Manager/Analyst in risk and safety as well as for Governance and Architecture of Data + IT/OT; Certified Data Scientist. 30+ years of international experience in risk and process safety, security, and infrastructure development. jcangarita@imsglobal1.com

Maria F. Angarita Cardona, Healthcare/Medical Devices leader. Physician (Universidad Libre); Certified Lead auditor in quality, safety, and security for medical Assets and Devices, Information Security, IT/OT/IoT Cybersecurity, Cloud Security, Privacy, and IT/OT Services; Certified Lead Manager in risk. 5+ years of experience in risk and security. mfangarita@imsglobal1.com

Sara C. Angarita Cardona, Biological & Agricultural engineering (Texas A&M University) student, Cybersecurity Bachelor+ program (Texas A&M Engineering Experiment Station), Business Minor, and Data Analytics track. Intern at IMS Global. Lead auditor in Information Security, IT/OT/IoT Cybersecurity, Food safety, HSEQ; Lead manager in compliance and Lead analyst in HACCP and risk. saraangaritac@tamu.edu

Abstract

When it comes to effectively protecting our information, information technologies (IT) as well as critical operational technologies (OT) and infrastructures, including the devices connected to the internet (IoT) and Industrial Automation and Control Systems (IACS/ICS), there is a worrying gap between current security and operational practices and the broad landscape of legal, regulatory and technical requirements framework. That framework sets common ground to identify and understand (i) the assets involved (information, IT/OT/IoT technology, and infrastructure) and how to identify, analyze, characterize, and protect them; (ii) the risks faced given the nature of the assets involved, the evolving complexity of cyber threats and dynamism of the ecosystem of cyber adversaries; and (iii) the minimum control solution expected at a technical, technological, and management level; to maintain and improve quality, safety, security, performance, resilience, and compliance. In many cases, organizations just use a technical reference, generally because they are comfortable with it, but that approach leaves aside the bigger picture of laws, regulations and technical standards and guidelines that are applicable and should add/protect value in the organization and in general in our critical infrastructure and our national security. The concept of compliance is often seen separately and can be perceived as a problem when it comes to audits or evaluations done by regulators such as CISA or TSA, or when a cyber incident occurs, in which DHS or the FBI will intervene.

Introduction

According to the European Commission[1] - through its Joint Research Centre - and the International Chamber of Commerce[2] in its first cybersecurity brief, the annual global cost of cyber-crime was $6.2 trillion in 2020, double that of 2015, the largest transfer and loss of wealth in history - at that time. In 2023 this problem exceeded $8 trillion and is projected to reach $10.5 trillion in 2025[3] and $14 trillion in 2028[4]. "*To maintain our competitive advantage, the United States must continue to harness its unmatched strengths - its private capital markets, a national security supplier base, and vibrant technological ecosystem*" from the U.S. Secretary of Defense[5].

Cybersecurity risk is no longer just an IT or OT/IoT issue – it really is a highly critical situation that directly influences the health of any organization and its stakeholders as well as its products, services, processes, and business environment. The domino effect does not end when the cyber-incident has been managed (e.g. ransomware) and the operations are restored[6]. Cyber-incidents can harm a brand's reputation, its financials, put our lives in danger, and affect our entire way of life. That is why - today - it is necessary to talk about cybersecurity risks as business risks.

The government of the United States and its agencies and departments understand that challenge. "*Keeping nation's critical infrastructure secure is important to our national security. Critical infrastructure spans everything from healthcare, water and education to chemical, transportation, and energy systems. It is interdependent with other critical infrastructure and supporting systems with all the essential services and critical functions that keep our country and our economy running. America's national security & economic prosperity are increasingly dependent upon critical infrastructure that is at risk from a variety of hazards, including both physical and cyber. Critical infrastructure security and resilience requires a clear understanding of the risks we face and a whole-of-community effort that involves partnership between public, private, and non-profit sectors*" from the Cybersecurity & Infrastructure Security Agency (CISA), November, 2022[7]

This article presents how a suitable understanding of compliance can add value to organizations of any kind, in special in the industrial world, by exploring key concepts, examples, common mistakes and misunderstandings, and showing technical recommendations to not just "comply" but, instead, learn how compliance management can be used as a positive factor for the advance of effective, efficient, quality, competitive, and safe operations. This paper relies heavily on technical concepts, definitions, and approaches from international technical standards and widely recognized technical centers as well as those established in the United States law and regulations.

---

[1] The European Commission. https://publications.jrc.ec.europa.eu/repository/handle/JRC121051
[2] International Chamber of Commerce, https://iccwbo.org/news-publications/policies-reports/icc-cybersecurity-issue-brief-1/
[3] Cybersecurity ventures. https://cybersecurityventures.com/top-5-cybersecurity-facts-figures-predictions-and-statistics-for-2021-to-2025/
[4] Statista. Cybercrime (historic and projected data). https://www.statista.com/forecasts/1280009/cost-cybercrime-worldwide
[5] DoD. https://www.defense.gov/News/Releases/Release/Article/3543777/department-of-defense-and-small-business-administration-roll-out-the-small-busi/
[6] Forbes. Assessing The Correlation Between Cyber Risk And Business Risk. https://www.forbes.com/sites/forbestechcouncil/2023/06/07/assessing-the-correlation-between-cyber-risk-and-business-risk/?sh=111611b44e7d
[7] CISA. November 2022 – Infrastructure Security Month. https://www.cisa.gov/sites/default/files/publications/ISM_2022_Toolkit_October%202022_Final-with-Cover_508c.pdf

The technological landscape

A compliance management analysis requires considering which are the security domains: There are six domains of security: (i) Information Security; (ii) Information Technology (IT); (iii) Operational Technology (OT) + Industrial Automation and Control Systems (IACS/ICS); (iv) Internet of Things (IoT) and Industrial Internet of Things (IIoT); (v) Cloud (CT); and (vi) Artificial Intelligence (AI). Information, IT, Cloud & AI domains require critical information infrastructure protection (CIIP) whereas OT+IACS/ICS & IoT-IIoT a critical infrastructure protection (CIP) approach. Institutions such as ISO, IEC, ISA, NIST, NERC, among others, address the specifics involved in CIIP and CIP, as well as different perspectives for their development.

In terms of critical infrastructure, the Cybersecurity & Infrastructure Security Agency (CISA) has defined sixteen critical sectors[8] "*whose assets, systems, and networks, whether physical or virtual, are considered so vital to the United States that their incapacitation or destruction would have a weakening effect on security, national economic security, national public health or safety, or any combination thereof*" following the Presidential Policy Directive - Critical Infrastructure Security and Resilience PPD-21[9]. Those sectors, and the corresponding agencies in charge of regulation and control in each case, are shown in the following image[10]. CISA, TSA, DOE, USDA, HHS, USCG and EPA are the main actors for operational technologies (OT).

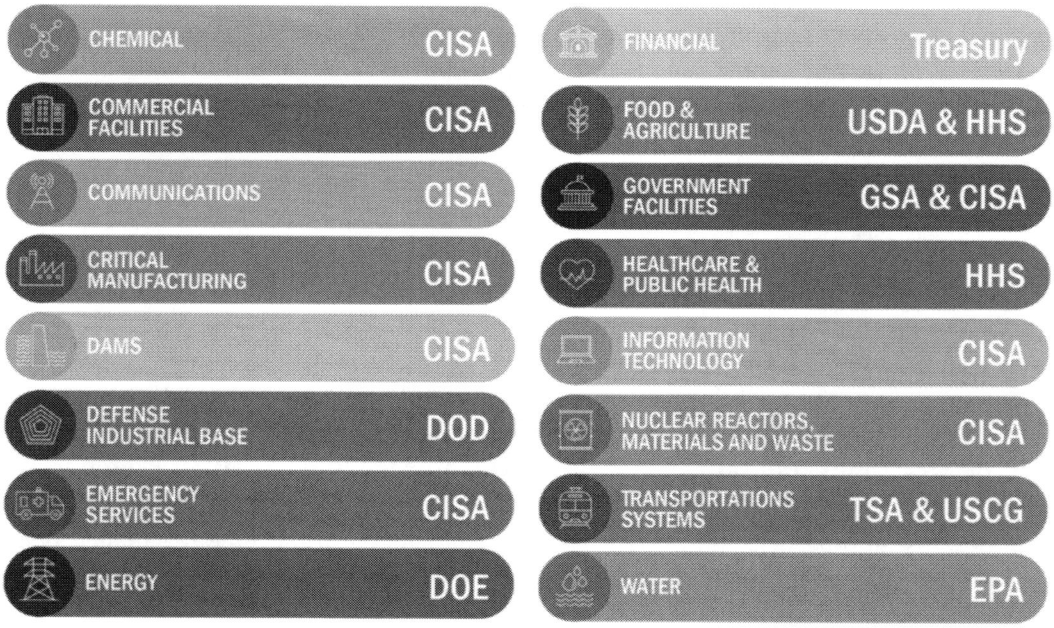

Fig. 1. Critical sectors and its agency(ies)

[8] CISA. Critical Infrastructure Sectors. https://www.cisa.gov/topics/critical-infrastructure-security-and-resilience/critical-infrastructure-sectors
[9] https://obamawhitehouse.archives.gov/the-press-office/2013/02/12/presidential-policy-directive-critical-infrastructure-security-and-resil
[10] CISA. CISA overview and services. https://www.nist.gov/system/files/documents/2021/05/06/LundyCISA%20BriefingSpace%20Cyber%20Symposium%20Access%20for%20Start-Ups.pdf

Those critical sectors can be graphically arranged to see both worlds: IT (including cloud and AI domains) and OT + IoT in the other side, using as a differential factor the type of technology that each sector uses for its operation as well as the type of protection required (CIIP or CIP), having as a common factor the information assets, as is shown in Figure 2.

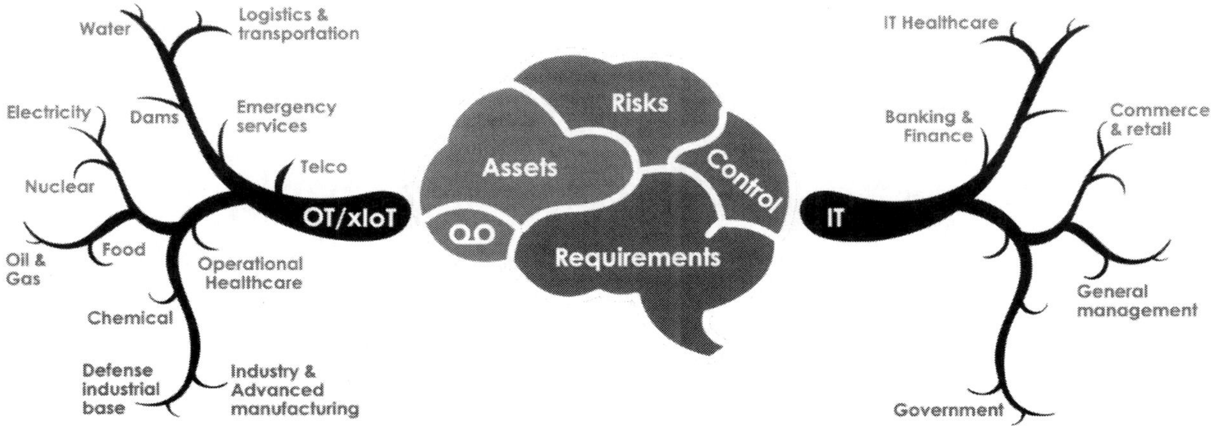

Fig 2. The IT+OT/IoT landscape (from the authors)

How requirements fit into the equation

ISO/IEC 15408 (and its adaptation in ISO/IEC 27032, previous version) establishes a conceptual relationship map between main elements in the overall information security landscape (Figure 3).

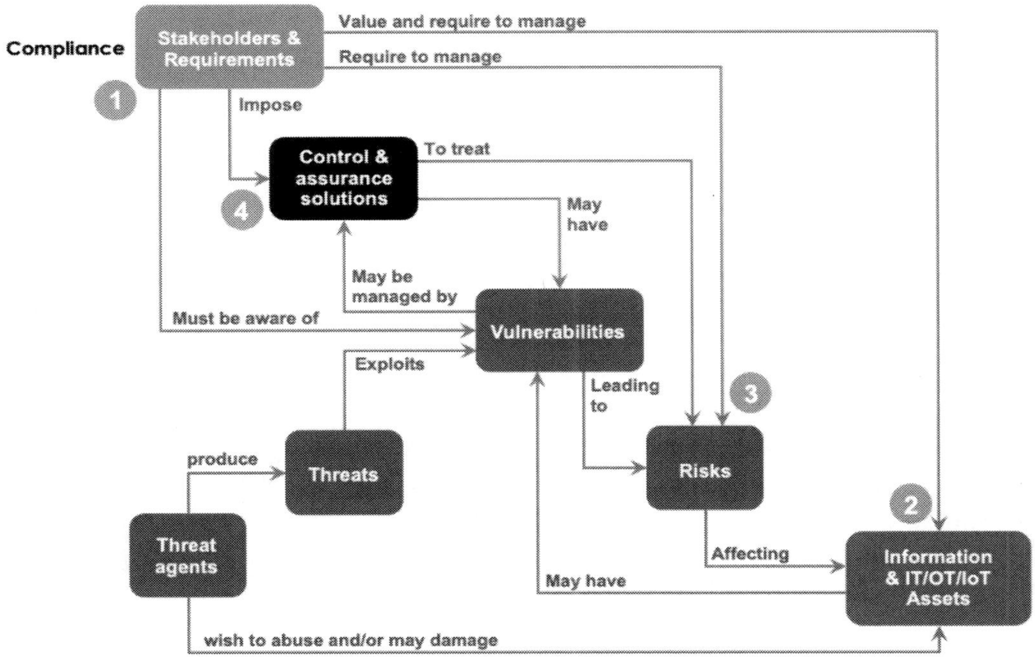

Fig 3. Concepts and relationships map in information security.

The following sequence of functional components by reorganizing the map in Figure 3:

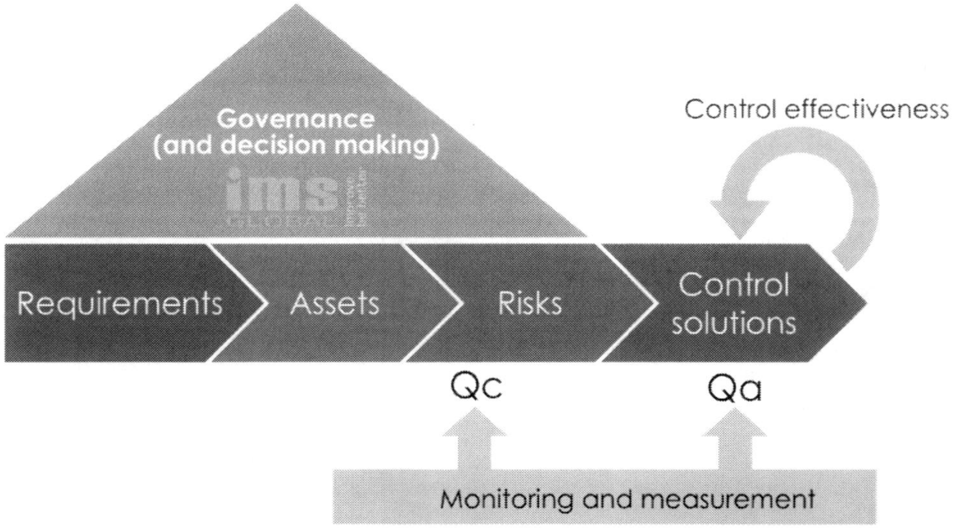

Fig. 4. Sequence of functional elements in information and technology security.

In this sequence, the requirements framework (legal, regulatory, and technical) is at the beginning because it can (i) define the context and scope of the (critical infrastructure) cybersecurity approach, program or management system (regardless of how it is required to be labeled); (ii) include rules to understand, classify, prioritize and improve information and technological assets during their life-cycle, including approaches for achieving high levels of maturity in asset management; (iii) contain directives about risk management practices (such as ISO 31000, ISO/IEC 27005, IEC 62443-3-2, Norsok Z-013 R3, NIST SP 800-30 R1, NIST SP 800-37 R2, CCPS-CPQRA, NERC, among others) or a hybridization of some of them (to the assets in the scope); (iv) define a common practice or minimal control set; and, transversely (v) define rules for the monitoring, measurement, and reporting activities as well as (vi) set the need for governance - aimed at maximizing the value generated by the assets - that has as inputs those listed above (as defined in the family of standards ISO/IEC 38500 for IT governance, and ISO/IEC 38505 for data governance, as well as the technical standards and guidelines ISO/IEC 27014 for security governance, ISO/IEC 38507 for artificial intelligence governance, NIST CSF 2.0, and OECD guidelines in the matter…).

This is when requirements must go from a table on a spreadsheet to become active elements that define an organization's approach toward a better achievement of objectives through the protection of its infrastructure, requiring that the organization has the ability to execute those requirements. The problem arises when **people start at the end**: applying a control solution without an adequate or sufficient knowledge of their assets or risks and - especially - the applicable requirements. The definition of control in ISO 31000:2018 "*measure that maintains and/or modifies risk*" raises the paradox of control without knowing what risks are "being controlled" (unfortunately, something common seen in our experience as auditors).

Requirements are, in fact, an active part of business process. The identification, determination, and application of legal, statutory, regulatory, technical, and contractual requirements is an elemental but necessary step to run processes properly as well as to provide adequate products and services (as is defined in the well-known standard in quality management, ISO 9001:2015). CISA, in this regard, from the perspective of security for critical industries, has questioned the phenomena of "vulnerability by design" in its secure by design publication[11].

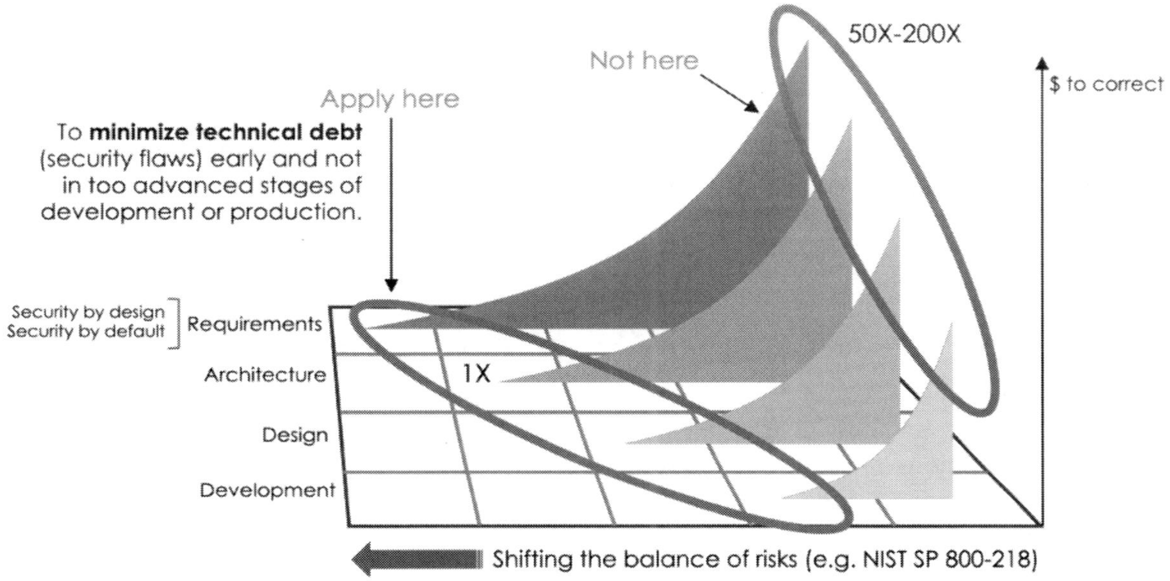

Fig. 5. Requirements in the acquisition (design & development or other mechanisms) cycle.

Requirements are also a core element for the best design, development, and upgrade of processes, systems, infrastructure, information and operational technology (including software, IACS/ICS, and information services), and the execution of projects, as well as for the provision of products and services, among other major activities in any kind of organization. Figure 5 shows how ignoring a requirement - in this case security by design and/or security by default - at the start of its design and development (or upgrade) process can cost up to 200 times the original amount of money. Some technical standards (such as NIST SP 800-218, in the case of software engineering) highlights the importance of shifting to the very start the application of requirements, minimizing technical debts (e.g. security flaws / vulnerabilities) and, as an effect, saving money.

The issue with compliance is not just sanctions or fines, it is the quality (or the degree to which a set of inherent characteristics of an object – including processes, products, and services – fulfills **requirements**, ISO 9000:2015) and security/safety itself of any process, technologies, products, and services in the organization, aggravated with the impact on reputation, finance, and lives.

[11] CISA. Secure by design. https://www.cisa.gov/sites/default/files/2023-10/SecureByDesign_1025_508c.pdf

Compliance as a matter of security - bad guys do not follow rules!

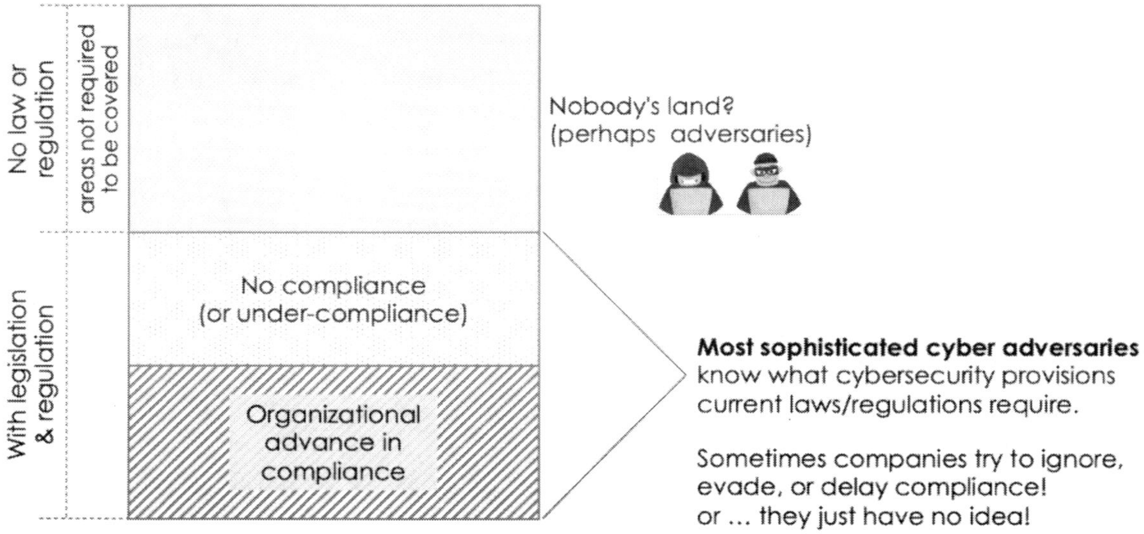

Fig. 6. Compliance vs regulated / non-regulated space

Laws and regulations are and will continue to be a strategic part of protecting people, systems, industries, environment, and society at large, by enforcing a higher standard of care by investing in good and proper technology, policies, processes, and people in organizations and their suppliers. This is an obstacle to the average bad guy and the criminal masses, and that is a good thing!

Assets, risk approach, and control schemas in critical industries are partly ruled, whose provisions are well **understood by advanced cyber-adversaries** and companies used to partially comply with the applicable requirements which leaves a very large area without protection.

Then, a clever adversary could be blocked only by (i) a risk-based approach to do smart investments in cybersecurity; (ii) an adequate understanding and prioritization of assets; (iii) a technical ability to prevent and respond earned through practice, preparation and learning, based on risk; (iv) a broad view of the complex landscape of cyber-adversaries and their doctrines; and (v) meet the baseline established in legal, regulatory and technical requirements. Solving this challenge is increasingly relevant because systems are becoming online faster than we can define requirements and protected them. Many of those systems manage our private information and other support our way of life.

So, what is compliance?

According to ISO 9000:2015, **compliance** can be understood as the state in which an organization meets all the requirements the organization has to comply with. In this context, an extra definition is also useful, **conformance or conformity**, related to those requirements that the organization has voluntarily chosen to comply with. Both terms are included in ISO 37301:2021 Compliance Management Systems (which replaces ISO 19600:2014).

A compliance and conformity managing tool, created by the authors of this paper, designated as the **visibility model of compliance**, expands the traditional Kelsen pyramid, proposed by the legal jurist and theorist Hans Kelsen in his book Pure Theory of Law (1934, Reine Rechtslehre, and 1960 its revised/expanded edition), and his disciple Adolf J. Merkl, as well as some key ideas from John Austin, an English legal theorist, and Herbert Hart, an English legal philosopher in his book The Concept of Law (1961). In simple words, this pyramid sets a "hierarchical structure of the legal order"[12] for rules from government, having the constitution at the highest level, followed by laws and regulations and considering that other normative elements can arise at the government level. In fact, the United States President "*can issue executive orders, which direct executive officers or clarify and further existing laws*"[13] and the agencies can publish "*guidance or policy statements, which further clarify how an agency understands and implements existing laws and regulations*"[14].

Fig. 7. Visibility model (created by the author)

In our model, at the highest level, in the compliance category (marked in red in Figure 7), are the requirements from the government. In general, laws are passed by both branches of Congress and signed by the President. Regulations are developed by executive branch agencies or administrative departments to interpretate the law and define how it will be implemented. Laws and regulations state requirements or prohibitions. For this analysis, these requirements were simplified to the set of laws and executive orders (E.O.) as well as regulations and guidance generated by agencies.

[12] Olechowski, T. (2018). Legal Hierarchies in the Works of Hans Kelsen and Adolf Julius Merkl. In: Müßig, U. (eds) Reconsidering Constitutional Formation II Decisive Constitutional Normativity. Studies in the History of Law and Justice, vol 12. Springer, Cham. https://doi.org/10.1007/978-3-319-73037-0_9

[13] The White House. The executive Branch. https://www.whitehouse.gov/about-the-white-house/our-government/the-executive-branch/#:~:text=The%20President%20can%20issue%20executive,and%20clemencies%20for%20federal%20crimes.

[14] U.S. Department of Health and Human Service. Public Halth Emergency (PHE). Introduction to U.S. Law & Policy. https://www.phe.gov/s3/law/Pages/default.aspx

The next level is the set of organizational requirements, subordinated to the requirements from the government. They are in the conformity category (in blue) because the requirements set by the organizations are voluntary and can be modified by the organization itself.

Then, at the same level, there are technical requirements (technical standards and guidelines from ISO, ISA, IEC, NIST, API, among others) falling into the conformity category (since their adoption is voluntary unless ruled by government laws or regulations). As will be shown in some of the following examples, technical requirements - being conformance - can move into the realm of compliance through incorporation by reference "*A method of making a secondary document part of a primary document by including in the primary document a statement that the secondary document should be treated as if it were contained within the primary one.*"[15]

On the other hand, there are the contractual requirements (both those established by the purchasing organizations and those included in signed contracts) located in the compliance category under the concept of "Contract as Promise: A Theory of Contractual Obligation"[16].

Example 1: To understand how a law establishes a requirement, and the corresponding agency[17] (in this case the FDA) develops the resulting regulation, there is the following case. Law 117-328 December 29, 2022, General appropriations, Section 3305: Cybersecurity for Medical Devices, amended the FDA Act (21 U.S.C. 351 V.5) adding the Section 524B - Ensuring cybersecurity of devices. This law introduced the concept of (Medical) 'Cyber device' as a "*device that: Includes software validated, installed, or authorized by the sponsor as a device or in a device; Has the ability to connect to the internet; and Contains any such technological characteristics validated, installed, or authorized by the sponsor that could be vulnerable to cybersecurity threats*". This Law is being regulated by the U.S. Food and Drug Administration. On September 26, 2023, the FDA published the guidance "Cybersecurity in Medical Devices: Quality System Considerations and Premarket Submissions" which supplements the FDA document "Cybersecurity for Networked Medical Devices Containing Off-the-Shelf (OTS) Software" (2005). On March 13, 2024, The FDA issued the draft guidance "Select Updates for the Premarket Cybersecurity Guidance: Section 524B of the FD&C Act". FDA also hired MITRE to develop the report "Next Steps Toward Managing Legacy Medical Device Cybersecurity Risks" on Nov. 15, 2023[18]. This Law and the work of the FDA focus on addressing "*Threats and vulnerabilities cannot be eliminated and reducing cybersecurity risks is especially challenging. The health care environment is complex, and manufacturers, hospitals, and facilities must work together to manage cybersecurity risks*". In addition, the FDA established that manufacturers of cyber devices are required to submit cyber security information (for new submissions) starting March 29, 2023.

[15] Garner, B. Black's Law Dictionary, Abridged, 9th Ed. 2010.
[16] Michigan Law review. Contract as Promise: A Theory of Contractual Obligation. https://repository.law.umich.edu/cgi/viewcontent.cgi?article=4690&context=mlr#:~:text=In%20Contract%20as%20Promise%3A%20A,1).
[17] U.S. Food & Drug. Fundamentals. https://www.fda.gov/about-fda/fda-organization#:~:text=FDA%20is%20an%20agency%20within%20the%20Department%20of%20Health%20and%20Human%20Services.
[18] U.S. Food & Drug. Cybersecurity. https://www.fda.gov/medical-devices/digital-health-center-excellence/cybersecurity

The following selection shows some executive orders on cybersecurity for critical infrastructure:

- Executive Order 12829, January 6, 1993, National Industrial Security Program
- Executive Order 13636 Feb. 12, 2013, Improving Critical Infrastructure Cybersecurity
- Policy Directive–21 (PPD–21), Feb. 12, 2013, Critical Infrastructure Security & Resilience
- Executive Order 13800, May 11, 2017, Cybersecurity of Critical Infrastructure.
- Executive Order 13873, May 15, 2019, Securing Information & Communications Technology and Services Supply Chain
- Executive Order 13920, May 01, 2020, Securing United States Bulk-Power System
- Executive Order 14017, Feb 24, 2021, America's Supply Chains
- Executive Order 14028, May 12, 2021, Improving the Nation's Cybersecurity

Example 2: In this case an executive order leads to the development and application of a technical standard. The Executive Order 13636, Improving Critical Infrastructure Cybersecurity, from the White House, introduces a definition of "*Critical infrastructure: systems & assets, physical or virtual, so vital to the United States that the incapacity or destruction of such systems and assets would have a debilitating impact on security, national economic security, national public health/safety, or any combination*". In addition to establishing guidelines about the "Voluntary Critical Infrastructure Cybersecurity Program" and the "Identification of Critical Infrastructure at Greatest Risk", this Executive Order also mandates the development of a "Baseline Framework to Reduce Cyber Risk to Critical Infrastructure" motivating the further development of the NIST Cybersecurity Framework (now in its version 2.0, published in February 2024).

Example 3: On the other hand, this third example illustrates the convergence between a law and an executive order resulting in a proposed regulation for acquisition (contractual requirement) and the application of some technical requirements (with NIST technical standards). The Public Law 116–207, December 4, 2020, Internet of Things Cybersecurity Improvement Act, and the Executive Order 14028, May 12, 2021, Improving Nation's Cybersecurity, proposed a Federal Acquisition Regulation (FAR) about "Standardizing Cybersecurity Requirements for Unclassified Federal Information Systems" FAR Case 2021-019, involving six special publications: NIST SP 800–82, Industrial Control Systems Security; NIST SP 800–213, IoT Device Cybersecurity; NIST SP 800–161, Cybersecurity Supply Chain Risk; NIST SP 800–53, Security & Privacy Controls; NIST SP 800–34, Contingency Planning Guide; and NIST SP 800–137, Continuous Monitoring (ISCM).

Maturity associated to cybersecurity and compliance

A term that has become common (and used incorrectly) is **maturity**. Although there are multiple approaches and scales that seek to "measure" maturity, always from a qualitative perspective, this paper uses as a formal definition of maturity the one established in ISO 9004:2018 "*Improvement, learning and innovation are embedded as routine activities across the whole organization and are evident in relationships with interested parties*". This concept is located at the highest level of maturity for subclause 11.1 "Improvement, learning, and innovation in such technical standard".

Another key technical reference is ISO/IEC 21827 which defines the ISO version of the Capability Maturity Model for Systems Security Engineering. Two important concepts are process capability which is "*an organization's potential; a range within which an organization is expected to perform*" and process performance as a "*measure of actual results on a particular project that may or may not fall within the range*". Maturity is addressed also in terms of how "*a specific process is explicitly defined, managed, measured, controlled, and effective*" implying "*the consistency with which it is applied*", that is, how the requirements are fulfilled. In this case, this model covers the processes used to perform the tasks of the specified domain (security engineering). It is important to note that ISO/IEC 15288 has very similar concepts in this topic.

In the case of maturity for a network of suppliers (and contractual agreements), there is the example of Cybersecurity Maturity Model Certification (CMMC) program from the Department of Defense (DoD)[19], aligned to the DoD's information security requirements for the Defense Industrial Base (DIB) partners, with three features: Tiered model, Assessment requirement, and Implementation through contracts. For the new version of the CMMC (2.0), the DoD will pursue rulemaking both in Part 32 of the Code of Federal Regulations (C.F.R.) as well as in the Defense Federal Acquisition Regulation Supplement (DFARS) in Part 48 of the C.F.R.

As for NIST, specifically speaking about cybersecurity and its use in critical infrastructure, there is the new version 2.0 (2024) which has four levels (1 - partial, 2 - risk-informed, 3 - repeatable, 4 - adaptive) in two categories: cybersecurity risk governance and cybersecurity risk management.

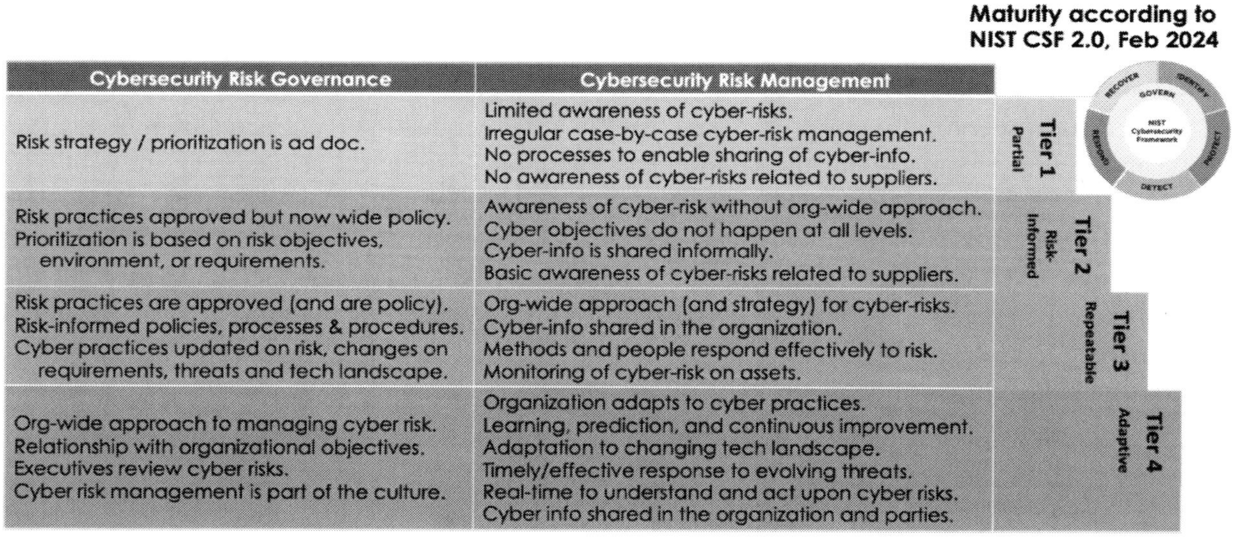

Fig. 8. New model of maturity in cybersecurity contained in the new version of NIST CSF 2.0 (published in February 2024)

---

[19] Department of Defense. Chief Information Officer. The Cybersecurity Maturity Model Certification (CMMC) program. https://dodcio.defense.gov/CMMC/About/#:~:text=The%20Cybersecurity%20Maturity%20Model%20Certification%20(CMMC)%20program%20is%20aligned%20to,with%20its%20contractors%20and%20subcontractors.

Finally, ISA/IEC technical standards for security for industrial automation and control systems IEC 62443-2-4 Security program requirements for IACS service providers and IEC 62443-4-1 Secure product development lifecycle requirements, at the same clause number 4.2 Maturity model, define a scale of Maturity levels, as is described in the following integrated table (compared vs CMM and ISO/IEC 15504-2 technical references):

Level	ISO/IEC 21827		IEC 62443	Short description
	SSE-CMM	ISO/IEC 15504-2		
1	Performed informally	Performed process	Initial	Product development and services supply is ad-hoc and undocumented (or partially) documented; consistency across projects and processes repeatability is not possible.
2	Planned and Tracked	Managed process	Managed	Product development and services supply can be done according to written policies (and objectives), its personal is competent; practices are repeatable (even under stress) and execution is according to planning.
3	Well defined	Established process	Defined (Practiced)	Performance is repeatable in the supplier's organization; process has been practiced.
4	Quantitatively Controlled	Predictable process	Improving	By using suitable process metrics, product suppliers and services providers can control the effectiveness and performance of the product/service and demonstrate continuous improvement.
5	Continuously Improving	Optimizing process		

Other key aspects of compliance management

* Relationship: Requirements are related to each other and emerge from (and depend on) of a higher hierarchical nature.

* Temporality: Requirements are created and approved at specific moments, their starting time of application or transition as well as its reporting are usually different, and they can have a finite life or be replaced in the future (when repealed) or lose their effect (when annulled).

* Granularity: It happens that not the entire body of a requirement, whether legal or technical, could be applicable in its entirety. It is necessary to determine which are the applicable elements and their relationship with elements of other standards (also at the same level of granularity).

* Jurisdiction: In countries with federal systems, or companies with complex and global structures, or when the operation involves several states and/or countries, the jurisdictions involved, and the issue of requirements collision must be observed.

We need to act now, avoiding the classical common excuses

- As long as the government or my clients don't require it, I won't do it.
- There are no rules on that subject, where do you get that from?
- As long as I comply with the XYZ guide, I have everything covered.
- Enforcing compliance damages the company, it only wastes money.
- I have a lot of technological toys, that's enough!
- We run checklists ourselves and we know that we are fine.
- Auditing and/or evaluating compliance is not part of the business.
- In self-diagnoses, we always do well.

That common excuses (and other) that authors have heard in auditing, training and even consulting, denotes a disconnection between the relevance of requirements management for the achievement of business objectives, the control and assurance of quality as well as security/safety of processes, products, and services, an overall effectiveness and efficiency, and the role of the organization in the security and resilience of its corresponding industrial critical sector as part of national security. It is still thought that requirements are only a problem when presenting reports or being evaluated, but it is not seen as part of the business and its strategy.

Conclusions

- Requirements are key to acquiring assets and ensure performance throughout its life cycle.
- Requirements set how to understand assets and risks, and as a result, control solutions to protect assets and ensure its expected function.
- If organizations do not - at least - comply with what is regulated, organizations cannot expect an effective protection & response against cyber threats.
- Requirements are central for the governance, use, and protection of assets.
- Early compliance minimizes technical debt and saves money.
- Verification criteria must be based on the requirements and risks related to the assets.

Upcoming developments

A future article will present our general methodological framework for comprehensive compliance management, applied to critical industries, strongly based on technical standards and aligned with relevant laws and regulations in the United States. Likewise, IMS Global has the TITAN GRC/IRM technology which implements the aforementioned reference methodological framework in its core compliance component. This development is designed to streamline, simplify and accelerate the processes of identification, application, evaluation, reporting, and improvement of requirements (both from government origin and technical sources) in critical industries, moving from a model of compliance "because it is mandatory" to one in which "requirements are value-generating tools".

68c Unlocking the Potential of Digital Technology to Enhance Process Safety in the Chemical Industry Keeping Security As a Top Priority.

Charles Fialkowski, Marco Ayala

In today's rapidly evolving technological landscape, the chemical industry is witnessing a paradigm shift towards digital transformation. This white paper explores the potential of digital technology to enhance process safety in the chemical industry while maintaining stringent security measures. By leveraging digitization, chemical plants can streamline operations, improve efficiency, and mitigate risks associated with safety instrumented systems (SIS). This paper delves into the key benefits and challenges of implementing digitization in safety systems, with a focus on maintaining robust security protocols. It also highlights best practices and case studies that demonstrate successful integration of digital technologies with safety instrumented systems, ensuring the highest level of safety and security in the chemical industry.

89a How to Develop Quickly New Kinetic Models By Including a Priori Information (Part I): Transfer Learning from Previous Catalyst Generations to New One

Per Julian Becker, Loïc Iapteff, Benoit Celse

How to develop quickly new kinetic models by including a priori information (Part I): Transfer Learning from previous catalyst generations to new one

Per Julian Becker[1], Loic Iapteff[1], and Benoit Celse[1]

1 IFP Energies Nouvelles, Rond-Point de l'Echangeur de Solaize, 69360 Solaize, France

**E-mail:* per.becker@ifpen.fr

HIGHLIGHTS

- Kinetic model parameter estimations with Monte-Carlo Markov Chains for new catalyst generation.
- Transfer Learning between catalyst generations by adding prior distribution of model parameters.
- Inclusion of hard & soft constraints in order to respect experimental rankings.

INTRODUCTION

Catalytic processes are immensely important to the chemical industry, with more than 95% of products using a catalyst at some part of the manufacturing process. Maximizing of product yields (and/or quality) while minimizing of costs, particularly in terms of energy consumption is becoming increasingly important considering the current global economic context of rising energy cost and requirements of reducing CO2 emissions. That is why, new catalyst generation are developed periodically to improve activity or selectivity.

Predictive reactor models are required for process design optimization, which often implies some extent of extrapolation beyond the range of available experimental data. Models must therefore be not only precise, i.e. within the acceptable range for the calibration and validation datasets, but also robust, i.e. avoiding aberrant model predictions when extrapolating. This is particularly challenging when the number of experimental data points limited and/or characterization of the reaction network is difficult due to the complexity of the feedstocks. For example, this is the case for VGO hydrocracking, where expensive and time-consuming pilot plant tests are required and the feedstocks are complex mixtures with more than 100,000s of hydrocarbon species. Typically, between 50 and 100 experimental points, with a wide variability of feedstocks (around 10) of different nature and origin are required for a good model, which is the case for mature catalysts. Model development for newly developed catalyst is particularly challenging as initially only few pilot plant points with a limited number of feeds are available. The scope of this study is then to improve the new catalyst kinetic model using the previous generations.

This work explores a methodology to include knowledge from previous catalyst generation(s) in a model for a newly developed catalyst via Bayesian Transfer Learning (TL) using Monte-Carlo Markov Chains (MCMC) [1, 2, 3] by applying a prior probability on the model parameters.

MATERIALS & METHODS

The example chosen to illustrate the technique is a Continuous Lumping (CL) model of the True Boiling Point (TBP) distribution with a single family and two reactions (cracking & isomerization); the Hydro De-Nitrogenation (HDN) reaction is solved simultaneously to account to the inhibition effect of organic nitrogen [4]. This formulation was found to produce robust, predictive models.

While machine learning methods have been used to model hydrocracking processes [2, 5], such models are generally only useful for process control, where the feeds and operating conditions remain in a narrow range and large datasets from industrial operating data is available. Purely data-driven modelling approaches, while precise, are not predictive and generally perform poorly in extrapolation.

The Bayesian Theorem gives the posterior distribution of the model parameters in terms of a prior distribution [1].

Where is the matrix of the observations (i.e. feed properties and operating conditions) and y is the matrix of output of the model, here standard cut yields and TBP distribution. The prior distribution is taken as the parameter distribution of robust model for a previous catalyst generation with a large number of available pilot plant points. While a closed form exists [1] for simple models, such as linear models, this is not the case for the mode complex CL model. The MCMC algorithm, which consists of randomly varying the parameters and applying a likelihood function, and optionally a prior, to determine whether to keep or discard the parameter variation, can be used in these cases. The prior distribution was obtained by running the MCMC algorithm on the large, mature, source dataset. This prior was then used for running the algorithm on the target dataset. The variance, or g-value, determines the adherence of the model to the value of the source parameters. The optimal g-value can be determined by performing cross-validation.

RESULTS & DISCUSSIONS

The methodology was applied to a source dataset of a previous catalyst with 91 points and 22 feeds and a target dataset from a new catalyst with 62 points and 14 feeds. A stacking test (cata1+cata2 and cata2+cata1) was used for model validation. The precision of the transferred model on the calibration and validation tests was found to be comparable to a model calibrated using a standard gradient descent algorithm without prior (transfer learning & imposed catalyst rankings); the expected catalyst rankings, in terms of diesel yield were however not respected in this case. The transferred model respects the catalyst rankings. The gradient-descent model clearly suffers from over-fitting. The MCMC algorithm can be used to produce more robust models with fewer experimental points by including information from previous catalyst generations and high-throughput test with are traditionally not used for model development.

KEYWORDS

Monte-Carlo Markov Chains, Parameter Identification, Continuous Lumping Model, Bayesian Statistics, Vacuum Gas Oil Hydrocracking

BIBLIOGRAPHY

1. Iapteff, 2022, Transfer Learning for Smart Predictive Analytics, PhD thesis
2. Tsung, et al., 2018, Statistical transfer learningn: A review and some extensions to statistical process control. Quality Engineering, pages 115-128
3. Iapteff, J. Jacques, M. Rolland, and B. Celse, 2021, Reducing the number of experiments

required for modelling the hydrocracking process with kriging through bayesian transfer learning. Journal of the Royal Statistical Society: Series C (Applied Statistics).

4. J. Becker, B. Celse, D. Guillaume, V. Costa, L. Bertier, E. Guillon, and G. Pirngruber, 2016, A continuous lumping model for hydrocracking on a zeolite catalysts: model development and parameter identification, Fuel, pages 73–82.
5. Pang, et al., 2024, Data-driven prediction of product yields and control framework of hydrocracking unit. Chemical Engineering Science, 1193386
6. Bouveyron and J. Jacques, 2010, Adaptive linear models for regression: improving prediction when population has changed, Pattern Recognition Letters, pages 2237–2247.
7. P. Robert et al., 2007, The Bayesian choice: from decision-theoretic foundations to computational implementation, volume 2. Springer.

89b How to Develop Quickly New Kinetic Models By Including a Priori Information (Part II): Transfer Learning from Fossil Feedstocks to Waste Plastic Pyrolysis Oil

Warumporn Pejpichestakul, Per Julian Becker, Benoit Celse

How to develop quickly new kinetic models by including a priori information (Part II): Transfer Learning from Fossil Feedstocks to Waste Plastic Pyrolysis Oil

Warumporn PEJPICHESTAKUL, Per Julian BECKER, Benoit CELSE

[a]IFP Energies nouvelles, Rond-point de l'échangeur de Solaize, BP 3, 69360 Solaize, France

warumporn.pejpichestakul@ifpen.fr

Abstract

The need to address plastic waste management and reduce reliance on fossil feedstocks in the petrochemical industry has sparked sustainable solutions. One such solution is the upgrade of waste plastic pyrolysis oil (py-oil) through hydroprocessing, which can utilize the existing facilities in the refinery and provide a means to transform the low-quality and impure py-oil into more valuable products that can be used as chemical feedstocks for the steam cracker to produce new plastics. However, this process is challenging due to contaminants, varying polymer compositions in plastic waste, and pyrolysis conditions [1].

Compared to extensive datasets available for conventional fossil feedstocks, experimental studies on hydroprocessing py-oil are limited. Generating experimental data is extremely time-consuming and expensive. It would therefore be beneficial to transfer knowledge from fossil hydroprocessing into models for the new feedstock(s). One promising approach to bridge the domain gap between these feedstocks is the transfer learning, a data-driven technique that leverages existing data and predictive models from fossil feedstocks to fine-tune models for py-oil. Transfer learning has shown success in enhancing kinetic and product property models for hydroprocessing fossil fuels [2–4].

This work focuses on developing product property and kinetic models for the hydroprocessing of waste plastic py-oil by applying transfer learning based on the Monte-Carlo Markov-Chains (MCMC) approach. The product property models are generally linear correlations, while the kinetic model is a more complex continuous lumping model [5]. We compare the model obtained from transfer learning (fossil to py-oil) with those developed solely from py-oil data, both with the same set of features. Several criteria are used to evaluate the models, which are accuracy and robustness. The exercise of including a new feature from the new domain (py-oil feed) is carried out on the product property model to evaluate the adaptability of the current transfer learning method, which is developed for to same feature space (homogeneous transfer learning), to the different feature space (heterogeneous transfer learning). Both sets of models perform well on the relatively small calibration dataset, but transfer learning leads to more robust results. We particularly emphasize the models' ability to predict outcomes with different py-oil feedstocks (not seen during calibration), thus ensuring their applicability across a range of scenarios.

KEYWORDS

Monte-Carlo Markov Chains, Bayesian Statistics, Pyrolysis Oil; Waste Plastics; Hydroprocessing

References

[1] M. Kusenberg, A. Eschenbacher, M.R. Djokic, A. Zayoud, K. Ragaert, S. de Meester, K.M. van Geem, Opportunities and challenges for the application of post-consumer plastic waste pyrolysis oils as steam cracker feedstocks: To decontaminate or not to decontaminate?, Waste management (New York, N.Y.) 138 (2022) 83–115.

[2] P.J. Becker, L. Iapteff, B. Celse, in: A.C. Kokossis, M.C. Georgiadis, E. Pistikopoulos (Eds.), 33rd European Symposium on Computer Aided Process Engineering, Elsevier, 2023, pp. 1053–1058.

[3] L. Iapteff, J. Jacques, M. Rolland, B. Celse, Reducing the Number of Experiments Required for Modelling the Hydrocracking Process with Kriging Through Bayesian Transfer Learning, J. R. Stat. Soc. Ser. C. Appl. Stat. 70 (2021) 1344–1364.

[4] L. Iapteff, J. Jacques, Benoît Celse, V. Costa (Eds.), Bayesian Transfer Learning to Improve Predictive Performance of an ODE-Based Kinetic Model, 2022.

[5] P.J. Becker, B. Celse, D. Guillaume, V. Costa, L. Bertier, E. Guillon, G. Pirngruber, A continuous lumping model for hydrocracking on a zeolite catalysts: model development and parameter identification, Fuel 164 (2016) 73–82.

89c Innovative, Non-Proprietary Condition Monitoring and Analytics Solution for Specific Reciprocating Compressor Asset Reliability Requirements

Anthony Fregosi, Michael Hastings, Eduardo Hernandez

Reciprocating compressors play a critical role in a number of industries, but require more maintenance than a comparable turbo-compressor. An effective condition monitoring and diagnostics strategy therefore plays an extremely important role for reducing downtime and life cycle costs of these machines while increasing reliability, safety, integrity and efficiency. There are a number of monitoring techniques that are available for detecting wear, component failure, leaks and other faults in reciprocating compressors that have been developed over the years. This case study presents an innovative method for implementing some of these techniques in an enterprise-wide data historian for optimizing the monitoring strategy.

191a Accelerating Viscoelastic Foam Development Via Descriptor-Based Experimental Design and Modeling

Ana Nunez, Fabio D'Ottaviano, Xinjie Tong, Christopher Thiede, Morgan Springs

The progress in viscoelastic foam technology garners specific attention within the realm of flexible slabstock foam, especially when considering its applications within the bedding industry. The slow responsiveness, low resiliency, and desired weight distribution properties of viscoelastic foam make it essential in the development of sleep products. The desired formulation is usually defined by targeted ranges of performance specifications and the preference and availability of raw materials.

Various formulation approaches exist for viscoelastic technology development, and studies have shown how independent variables of raw materials affect viscoelastic foam properties. However, the relationships among the codependent variables and their synergistic effect on foam properties remain unknown. The lack of systematic knowledge of the descriptor-property relationship of viscoelastic foam leads to inefficiencies in conventional approaches where multiple formulation cycles are usually needed to achieve the desired properties.

Over the past few years, Dow Polyurethanes has invested substantial resources in advancing digitalization and predictive capabilities via its Predictive Intelligence program. The use of experimental design and statistical models within the chemical domain has empowered researchers and manufacturers to develop new products, optimize production processes, and improve sustainability in a more efficient way.

This study outlines a systematic approach for learning descriptor-property relationships of viscoelastic flexible foam. A descriptor-based experimental design was carried out and empirical models were built to capture the variability in foam properties. The model was validated based on the candidate formulations with predicted foam properties in a specific area of interest. With the implementation of the model, formulations with targeted properties were obtained in a significantly reduced formulation cycle compared to conventional formulation approaches. This work shows the contribution of statistical tools in directing and accelerating the development of intricate systems in the Polyurethane industry.

191b Enhancing Data Science in Agrichemical Manufacturing: Challenges and Solutions

Prabhas Hundi, Pritam Chanda, David Kelly, James Morden, Julie Fleming, Nic Ellis, Christopher Shuster

In the modern agricultural chemical (Ag-Chem) manufacturing landscape, the integration of sensors and data logs into daily operations has ushered in an era of abundant data. This wealth of data presents a fertile ground for data science applications, with the potential to bring about substantial improvements in productivity, operational efficiency, and the resolution of unexplainable plant behaviors. Data science techniques, including machine learning predictions, anomaly detection, and root cause analysis, offer significant promise, provided one can effectively navigate the challenges posed by the unique nature of manufacturing data.

Manufacturing data in Ag-Chem poses multifaceted challenges that demand rigorous pre-processing and curation before it can be effectively leveraged for data science applications. The primary context behind these challenges lies in the fact that this data is not obtained from controlled experiments, but instead, is a by-product of plant operations which are primarily geared toward achieving a consistent routine. This results in limited availability and variability of data in features and target variables, which are essential components of well-structured data science problems. Consequently, data scientists confront challenges on two fronts. On one hand, there are specific data preparation challenges, including addressing data noise caused by inaccurate measurements, harmonizing data points distributed across multiple vessels and timestamps, and integrating features from multiple systems with unsynchronized time resolutions. On the other hand, broader challenges exist related to data availability and usefulness. These encompass the scarcity of data variability and the lack of contextual information about background changes within the plant, which can lead to a high quantity but low-quality dataset for deriving insights. Additionally, data can lack specific features due to the absence of corresponding sensors or routine sampling procedures, or the influential features are not controllable by the operators giving rise to the risk that the insights from data science studies may not yield actionable solutions.

This presentation aims to shed light on each of these challenges, highlighting the resulting disadvantages they impose on data science efforts within manufacturing. Moreover, it offers practical solutions and insights whenever possible, such as procedures for pre-processing data to circumvent the unintended misrepresentations that may lead to underperforming models or inaccurate insights. Real-world case studies from projects undertaken by the Advanced Analytics team in the Crop Protection Manufacturing division at Corteva Agriscience are presented to illustrate the impact of proper data collection and curation, highlighting the transformation of outcomes. The information highlighted in these studies helps in gaining a deeper understanding of the intricate challenges inherent in Ag-Chem manufacturing data and practical techniques to overcome these hurdles. The use of real-world case studies ensures that the insights and solutions are grounded in tangible experiences, making this presentation a valuable resource for data scientists, engineers, and professionals involved in manufacturing and data analysis.

108a Bridging Machine Learning and Adaptive Experimental Design: A Collaborative Paradigm for Accelerated Product Development

Joel Strickland

Abstract:

In the current landscape of product manufacturing, accelerated development stands paramount. At the confluence of collaboration and innovation lies a transformative approach that seamlessly integrates multiple knowledge domains, fostering an environment conducive to rapid product ideation, testing, and market release.

Intellegens, leveraging its expertise in handling sparse and noisy experimental data, presents a collaborative model. This paradigm marries the depth of subject-matter expertise with the agility of adaptive experimental design and the predictive prowess of machine learning, thereby championing a multi-faceted approach to product development.

Key highlights of our approach:

1. **Interdisciplinary Collaboration:** A synthesis of domain-specific knowledge and advanced data methodologies, driving a harmonized product development journey.

2. **Adaptive Experimental Design:** A dynamic approach to experimentation that iteratively refines based on preliminary data, optimizing for information richness and resource efficiency.

3. **Machine Learning Integration:** Harnessing predictive models to identify optimal experiment pathways, effectively navigating the vast design space and elucidating complex relationships.

4. **Impactful Case Studies:** Real-world examples from our collaborative endeavors across diverse sectors, illuminating the tangible benefits of accelerated product cycles and knowledge augmentation.

5. **Visionary Outlook:** Peering into the horizon of product development, emphasizing the sustained significance of collaboration and adaptability in shaping the industry's future.

Intellegens underscores the transformative potential of this collaborative paradigm, challenging traditional product development norms. We extend an invitation to industry peers to join this movement, collectively redefining the contours of efficient and innovative product creation.

129b Determining Ion Activity Coefficient Using Graph Convolutional Neural Networks in Ion-Exchange Membranes

Pegah Naghshnejad, Teslim Olayiwola, Gabriela Theis Marchan, Jose Romagnoli, Revati Kumar

Electrically driven ionic separations have held historical significance in water desalination and ultra-pure water production for applications in semiconductors, food, and pharmaceutical manufacturing. These electrically driven ionic separations encompass relying on ion-exchange membranes (IEMs). The chemistry and molecular structure of these IEMs significantly influence their ionic conductivity, permselectivity, and other transport properties, such as osmotic drag. Consequently, the chemistry and microstructure of IEMs exert a profound influence on these properties.

The partitioning of ions between polymeric ion-exchange membranes (IEMs) and the surrounding liquid is governed by the activity coefficients of the ions, which, in turn, have a significant impact on various ion transport processes within these membranes, notably conductivity. These coefficients are essential for comprehending phase equilibria, solubility, and chemical reactions, making them indispensable for a wide spectrum of industrial and scientific applications. In the past, estimating activity coefficients typically involved complex mathematical models and the collection of extensive experimental data.

This study introduces a computational framework designed to predict the activity coefficients of ions in charged Ion Exchange Membranes (IEMs). In recent years, machine learning (ML) has been integrated into materials design workflows as a complement to experiments and simulations to accelerate the discovery of a wide range of materials.[2] Graph Neural Networks (GNNs) have emerged as a powerful tool in the field of chemistry, offering an innovative approach to modeling and predicting diverse molecular properties, including the crucial domain of activity coefficients. Specifically, the framework utilizes Graph Convolutional Networks (GCN) to establish connections between the chemical structure of the polymer and the molecular-level attributes. This ultimately leads to the determination of macroscopic attributes, such as the activity coefficient, across a range of IEM materials, including random copolymer and block copolymer systems.

Data from our previous work based on the electrolyte molecular structure and composition alone with Molecular Dynamics simulations was used to train our model. Our dataset comprises a comprehensive combination of experimental and molecular dynamics (MD) data, which served as the foundational elements of the study conducted by Gallage Dona et al.[1] This method utilizes 137 different input variable categories derived from polymer characteristics, salt ions, the quantity of water molecules attached to ions, salt concentration, and twelve solvation descriptors obtained from MD simulations as the target variables.

The architecture encompasses a series of meticulously designed stages, each contributing to the model's exceptional performance. The initial phase of the architecture uses different data structures such as polymer structure and salt structure as an input to graph convolutional neural network. Following the graph-based transformation, the architecture leverages Convolutional Attention Networks (CAT) to seamlessly fuse the graph-based molecular representations with critical features such as salt concentration and Molecular Dynamics (MD) descriptors. The culmination of the architecture involves a meticulously crafted Feedforward Neural Network (FNN). This FNN component, driven by the consolidated data from the previous stages, serves as the predictive pinnacle, generating precise and insightful predictions for activity coefficients. Impressively, the ML-MD modeling strategy yielded highly accurate predictions of ion activity coefficients within IEMs.

References

[1] Hishara Keshani Gallage Dona, Teslim Olayiwola, Luis A. Briceno-Mena, Christopher G. Arges, Revati Kumar, and Jose A. Romagnoli, 'Title of the Paper,' Industrial & Engineering Chemistry Research, vol. 62, no. 24, pp. 9533-9548, 2023, doi: 10.1021/acs.iecr.3c00636.

[2] G. Bradford, J. Lopez, J. Ruza, M. A. Stolberg, R. Osterude, J. A. Johnson, R. Gomez-Bombarelli, and Y. Shao-Horn, 'Chemistry-Informed Machine Learning for Polymer Electrolyte Discovery,' ACS Cent. Sci., vol. 9, no. 2, pp. 206-216, Jan. 23, 2023, doi: 10.1021/acscentsci.2c01123.

108c Methodologies for the Inverse Design of Polyurethane Foam Materials

Daniel Christiansen, Aatish Pradhan, Sukrit Mukhopadhyay, Fabio Aguirre-Vargas

Informed materials design remains a foremost challenge for scientists and engineers. Having an ability to ideate a composition or process resulting in products with targeted specifications would be highly valuable in the chemical industry.

Recent uses of machine learning (ML) have enabled the connection of design and process to resultant material properties, a process known as *forward design*. A high-accuracy forward model enables the screening of potential material candidates *in silico* before spending resources to produce a formulation in the lab. The reverse of this process, going from property to formulation, called *inverse design*, results in a one-to-many challenge and is hindered by a typically discontinuous property-design space.

Here, we present approaches to inverse design to produce novel formulations having properties predicted to satisfy user-supplied design requirements. These strategies are demonstrated for polyurethane flexible foam materials utilizing real-world component and formulation data and expert-informed constraints.

129a Improved Understanding of Materials' Formulations and Design By Combining Multiscale Molecular Modeling with AI/ML Techniques

Jan-Willem Handgraaf, Ravindra Aglave, Panos Petris, Giulia Biffi

Data-driven modeling approaches continue to get more attention both in academia and industry. This is evident from the tsunami of papers published in the last decade on data driven modeling, including AI/ML techniques, and the success it actually has in industry when it comes to model accuracy and prediction, and the availability of relatively cheap computing power, while at the same time allowing companies to start making use of their large internal data pools.

In this work we discuss four industrial use cases. The first one deals with battery manufacturing where we simulate the processes to generate anode and cathode materials from slurry drying to calandering and we will showcase how to integrate these results in battery cell testing. [1]. The second case discuss the design of the battery separator material, which are typically polymer blends that experience non-llnear stress/strain behavior in typically small temperature range. We apply coarse-grained simulations techniques to be able to address the required time and length scales and show how to obtain the optimum polymer blend and corresponding material safetely specifications. Thirdly, we investigated by molecular simulation how to boost the performance of lithium ionic conductivity of solid state materials [2]. And finally, we showcase how multiscale modeling can be applied to the optimization of complex nutrient formulations for medical applications and how to relate it to key product characteristics like viscosity and texture.

When applicable we will apply AI/ML techniuqes to further enhance the robustness of the modeling in terms of predictive power and raw speed, where have used both in-house available methods and software tools from Google and Citrin Informatics [3].

[1] Microscopic investigation of the slurry drying process and binder migration in Li-ion battery anodes using multiscale-simulation methods, S Indrakumar, J Breitenbach, S Rudolf, JW Handgraaf, L Riewel, New Materials for Future Mobility 883 (Ingénieurs de l'Automobile), 35-40.

[2] Adeli et al., Boosting Solid-State Diffusivity and Conductivity in Lithium, Superionic Argyrodites by Halide SubstitutionAngew. Chem. 2019, 131, 8773 –8778.

[3] https://citrine.io/wp-content/uploads/2021/04/Case-Study-Panasonic-Technical.pdf

108b AI-Driven Materials Innovation for Industry 4.0: The PolymRize^{TM} Approach

Rishi Gurnani, Chiho Kim, Huan Tran, Rampi Ramprasad

The Materials Genome Initiative (MGI) and Industry 4.0 have heralded a sea change in the philosophy of materials and chemical formulations design. In an increasing number of applications, the successful deployment of novel materials and chemical formulations has benefited from the use of artificial intelligence (AI) coupled with experiment. Here, I will use the example of polymeric materials to describe the critical components of this transformative AI-based approach, including computational and experimental data generation and capture, polymer fingerprinting, algorithms for machine-learning based property prediction (polyGNN, polyBERT, etc.) and for the design of polymers meeting target property requirements (polyG2G, chemical variants, etc.), and finally, how prior physics knowledge can be woven into polymer informatics workflows. These efforts have led to PolymRize^{TM}, an accessible online platform for materials discovery and design, enabling faster and more cost-effective innovation. PolymRize^{TM} may be used with or without coding or specialized knowledge, making AI tools accessible to a wide range of users.

129c Machine Learning-Aided Design of Biodegradable Polymers

Chiho Kim, Huan Tran, Rishi Gurnani, Jessica LaLonde, Babetta L. Marrone, Ghanshyam Pilania, Rampi Ramprasad

In the pursuit of sustainable materials, biodegradable polymers have emerged as promising alternatives to traditional plastics, finding applications across diverse industries. Particularly, the degradation products of traditional plastic are of specific concern, and fully biodegradable and non-toxic plastic alternatives offer promising solutions to this ongoing challenge. However, a rational design approach to engineer biopolymers for degradation after their intended use still remains elusive. This is largely due to our inability to understand and model performance metrics (such as weight loss over time) capturing degradation behavior of these materials as a function of chemical, geometric and environmental factors.

As an exciting development in this direction, we have established predictive machine learning models which utilize physics-informed deep neural networks (NN) on previously established, manually curated experimental data that characterizes the weight loss behavior of fully biodegradable polyester copolymer samples in both water and soil natural environments. The models were applied in a series of experiments to predict the mass loss of approximately 10,000 structural and compositional variations of the 230 homo- and copolymers originally included in the original training set. The predicted mass loss for these candidate biodegradable polymers over a period of 365 days allowed us to identify novel copolymers with the potential to replace existing chemistries while matching property values to the desired performance metric ranges. This talk will discuss our findings and future directions, including the integration of critical properties, such as, thermal and mechanical characteristics, into the screening and design workflow and thereby bringing us one step closer to realizing the grand vision of a sustainable circular plastic economy.

109c The Use of AI Technology and Analytics to Improve the Process Efficiency in a 'Crystallizer' Unit Operation

Anthony Fregosi

After having attended a Webinar: AIChE (Fero Labs) – "Industry 4.0: 3 Ways to Accelerate Decarbonization while Cutting Costs". I saw great potential to use the concepts being developed by the presenters (Fero Labs) in one of our units.

Fero Labs uses AI technology to help manufacturers find and eliminate process inefficiencies.

109b Streamlining Digital Transformation: Discuss Approaches, Perspectives, and Success Stories!

Alexander Pfalzgraf

Digital Transformation has been around for more than 10 years so it is no longer a question if companies should transform digitally or not. It is time to review some key benefits that have been achieved, which will help you understand where to identify opportunities to transform.

When chemical companies are trying to advance their business, it is typically in the areas of productivity, efficiency, security, and sustainability. Optimizing production processes by using digital technologies such as AI, data analytics, and secure networks is common nowadays. But one challenge remains: which of these technologies are relevant to you and your company?

One way to gain insight of "the best digital fit for your company" is to assess where you currently stand and to measure the gap to the best-in-class performers in your industry. Analyzing these gaps will help to identify areas of improvement, and by closing these gaps you can ensure to stay relevant in a highly competitive environment. To demonstrate how this can be done, we will discuss a framework for assessing your plant and defining a vision of the plant of the future.

The format of this session will be a brief, introductory presentation with interactive elements in the first half, and an open discussion involving the audience in the second half.

During this presentation you will gain the following insights / participate in the following discussions:

- Hear about findings and success stories from multiple years of digital transformation projects in the chemical industry
- How to streamline a digital transformation concept for your company
- Participate in an open discussion about different approaches, perspectives, and success stories with your industry peers

109a Predictive Analytics Mega Implementation: Data Challenges and Successes

Hashim Al Hajji

Asset Management is a critical aspect of Energy Industry operations, which requires advanced and integrated digital solutions to enhance efficiency, reliability and safety of the assets and processes. However, implementing such solutions is not a trivial task, as it involves many challenges such as data quality, management of change, long-term support models and value capturing tools. In this paper, we present a novel solution that leverages predictive analytics to optimize the performance and lifecycle costs of hundreds of rotating equipment in a large-scale corporate wide deployment. The solution is based on a robust and innovative integrated system that supports the facility to achieve its goals by minimizing asset failures and operational upsets proactively. the solution also aims to empower the operation , engineering and maintenance staff with timely and strategic support from various SMEs within the organization as well as foster collaboration among different disciplines through a set of advanced and automated decision support systems. The focus of this paper is to showcase the successes and challenge of implementing predictive analytics solution trough a large-scale deployment such as improving asset utilization, reliability and availability, integrity, safety, and sustainability which enhances operator decisions and overcome data-related challenges. The strength of this deployment is that diversity of assets that has tackled, predictive models for over 5 types of rotating equipment were studied, prepared and tested to check the accuracy of prediction based on FMEA studies. Moreover, the impact of data quality, data availability, and adoption management were studied after this implementation. In this submission the aim is to showcase the lessons learned and best practices from Aramco's journey in this specific large-scale advanced analytics implementation.

130a Leveraging Digital and Factory Automation with Help of a Centrally Located Integrated Plant Operation Centre Outside Plant Premises.

Abhineet Raj, Naveen Kumar

Background: In line with Factories of Future, leveraging digital and factory automation, it is imperative to establish a near site unified command and control operation for Sintering & Raw material bedding and blending plant. With an emphasis on synergy, agile working, cross learning, faster and effective decision making and better coordination amongst all control stations, all six (06) control rooms is shifted to a centrally located integrated Plant Operation Centre.This transformation will be instrumental in future autonomous operation of Plant.

Uniqueness: Delinking Talent from Assets Due to the nature of the Iron Making value chain, factories are typically in remote locations, where the best of talent might be scarce and operational hazards are many. By separating the operations control from shopfloor processes, Tata Steel has laid the groundwork to ensure that geographical boundaries do not limit availability of talent, process, safety, environment and work can move anywhere than vice versa. With talent delinked from asset location, expertise is transferred remotely, ensuring rapid and efficient deployment of best practices even when new plants are set up. Operations, Safety and environment have also become safer with lesser people deployed near hazardous zones Technological enablement Real time data feed (voice, video & data overlay in unified layer) Fault Tolerant Architecture through distributed computing Edge Cluster and Cloud leveraged, digital twin and analytics enabled

Business Connect : It helped in overcoming from following shortcoming of traditional plant control (through separate control rooms):

a. Limited visibility to incoming raw material. The visibility is provided through periodic lab reports provided by scientific services. Real-time pile quality is not visible in Pellet Plant, SP1, SP2 control rooms.

b. Communication gaps regarding breakdown among different units (Pellet Plant, Sinter Plants & RMBB, and among the different units of Sinter plants & Material Despatch Control Room) leading to delayed action, thus productivity loss.

c. Decentralized operation planning among agglomerate plants leading to siloed decision-making.

d. Enhancing knowledge transfer and developing a multi-skilled pool of shared human resources among agglomerates plant.

e. Enhancing safety of field operators by digitally enabling the control rooms through CCTV footage in critical areas and reducing man-machine interactions.

Impact: Reduced number of incidents in dept Savings of 35cr per year Agile decision making, faster deployment Synergy and cross learning Benchmarking with all Sinter plants across India Achieved Best ever productivity & Stack emission. Reduction in Solid fuel rate by 5kg/tns over FY'22

130b Enabling Alive and Aware Digital Twins in Process Industries

Sushant Garud, Iftekar Karimi

The digital ecosystems of chemical and process industries are expanding with the rapid pace of digitalization. Several models are being developed using various digital twin technologies (first-principal models, machine learning models, simulation models, etc.) and native development software. Many Industry 4.0 technologies employ soft sensors and/or digital twins under the hood, which in turn depend on the (simulation) models. These models (digital assets) are typically developed by the domain experts using various software but utilized by plant personnel with diverse backgrounds and levels of training. Understanding and utilizing such models is not straightforward and demands significant time and attention from the user. Due to this, these models exist in silos and are shelved when not in use, rendering them stale or dormant (i.e., out of tuning) for dynamic operations. However, it is a time and resource consuming task to tune such dormant models before re-deploying them for plant data analysis. Moreover, due to the specialized nature of these models, their accessibility is limited to the expert users, creating a long turn-around time for operations analyses. A very real concern in the minds of industry operators is how to make sure that their huge investment in sophisticated digitalization technologies does not go to waste due to changes that their process and/or operations may experience. It would be desirable to overcome or ameliorate at least one of the above-described problems, or at least to provide a useful alternative.

Aleph Digital Technologies is developing a low-code software platform for managing generation and updating of digital twins of physical assets. A new paradigm for sustaining the digital twins developed using different technologies, birthed by different native software, and deployed at different scales is presented. The proposed approach facilitates central hosting and evolution of various digital twins through continuous assessment, identification of performance issues, and smart decision making for model calibration, training, and versioning. With this, the digital twins are kept alive and aware, in turn maintaining their accuracy and reliability. We have been actively working with various chemical and pharmaceutical plants to facilitate their sustainable digital ecosystems. One such example is a chemical process plant (CPP) based in India, let us call them CPP due to confidentiality reasons. CPP has embarked on a digitalization journey and wanted to adopt a low-code digital twin solution to accelerate and streamline their troubleshooting activities. In this work, we present a case study for one key problem that our proposed platform solved for CPP.

CPP was experiencing a lower-than-expected throughput for one of their distillation columns. Their in-house team of engineers tried to do manual troubleshooting analysis for months that temporarily helped them to increase throughput slightly. Due to this persistent issue, they decided to adopt our low-code, automated digital twin management platform. We began by building a high-fidelity first principles process digital twin for their system, followed by calibrating it to operational data. Next, we used the process twin to analyze the factors that could be responsible for low throughput. With this, our ML-based optimization engine utilized the process twin to determine the optimal process variables that can help achieve the desired throughput without affecting the product quality at the least cost. Our optimization suggested that there is a need to control reflux flow rate by installing a reflux drum that would reduce the reboiler duty by 5-7%, thus reducing energy and cost. It also highlighted the need to preheat the feed to higher temperatures by installing a preheater so that the column can give 15-20% higher throughput as the reboiler reaches its heat transfer limits. It also enabled operators to determine the optimal operating conditions under different feed conditions. Additionally, our software platform enabled continuous monitoring of product quality through process twins, thereby reducing the need to do frequent lab analyses. Overall, the deployed software platform and digital twin enables both domain experts and non-domain personnel to perform simulations, assess the impact of various operational decisions on product quality, energy cost, and emissions.

130c Combating Corrosion Under Insulation By Industrial Iot Monitoring

Stian Søvik, Eivind Vethe

Problem description

Corrosion under insulation, often known as CUI, has been one of the largest unsolved industry challenges for decades. Vast resources are spent on finding and combating the issue, while the threat of catastrophic failures remains for corroded areas that go undetected.

Production plants may contain hundreds of kilometers of piping systems to transport fluids or gases under pressure. Leakages must be avoided at all costs, due to the risk of disastrous explosions and severe pollution. Corrosion is a major contributor to this risk, as it tears down the wall thickness and reduces the strength of pipes. Several factors affect the risk and rate of corrosion, such as coating quality, material and surface temperature. However, wetting is the most important factor – without water, there is no corrosion.

A large portion of these pipes are insulated. The insulation is protected by metal cladding to keep water from intruding into the insulation. However, water tends to find its way through the cladding, for instance as rain water intrudes small openings in the cladding. As the cladding surrounds the insulation and process piping, it may hide the indications of any wetting or corrosion, appearing pristine from the outside. The insulation might also retain water. This increases the risk of corrosion going undetected and drives a tremendous effort in inspection and maintenance of insulated piping to combat CUI.

Current state of maintenance planning

To maintain the integrity of a process plant, necessary maintenance must be performed. As CUI slowly and covertly develops and eventually causes a leakage, one must have a strategy to detect corrosion or factors leading to it in time to mitigate the threat. Risk based inspection (RBI) methods assess the probability of factors that may lead to CUI and determine the priority and order of inspection based on an overall risk evaluation. Since CUI is a hidden threat, inspection often involves removal and replacement of cladding and insulation.

Operators have different RBI strategies and methods, but typical factors considered are process temperature, asset age, pipe material, coating type, and degree of wetting. Most of these factors are well known and can be obtained from documentation. However, while being one of the most important causes of CUI, wetting is highly unpredictable and variable, and its risk is often classified based on assumptions.

The uncertainty of the results of RBI assessments lead to overwhelming resources being spent on inspection and maintenance, and tons of replaced cladding and insulation wasted causing negative environmental impact. Inspection campaigns typically find that over 90% of the inspected areas are dry and unaffected by corrosion. At the same time, there is always a risk of corrosion appearing in the areas not prioritized for inspection as the location of wetting is highly unpredictable. All in all, the current methods of maintenance planning lead to high costs and remaining risks that could be improved by sufficient knowledge about the wetting.

Several attempts have been made to determine the degree of wetting to enable more precise and accurate risk assessments. Some methods rely on personnel operating large and expensive equipment in the process plants and have not shown satisfactory results. Other methods provide continuous monitoring of moisture but are costly and inefficient to install and operate. These experiences have shown that to safely and efficiently predict the risk of CUI, broad coverage of continuous wetting monitoring is required. This leads to the need for cost efficient and smart industrial internet of things (IIoT) sensors, and software solutions that can transform the vast amounts of raw measurements to actionable insights that enable optimized maintenance plans.

Technology

To enable cost efficient IIoT monitoring with dense coverage in large process plants, the sensor production cost is merely one factor. The total lifetime cost also consists of costs related to installation, provisioning, integration, and operation. The battery life determines how many years the costs can be split by.

Installation can be very costly, as piping systems often need scaffolding to access. Furthermore, wireless technology usually requires installation of gateways with power and network cabling throughout the site. To keep the installation costs down, the Fusion sensors are equipped with a fixation device allowing them to be plugged into the cladding in seconds. An optional self-drilling tool enables future robotic installation. More importantly, the solution eliminates the need for gateways as each sensor communicates directly with the CirruSense cloud through cellular 5G base stations that are usually already available in industrial areas.

The smart sensors are configured and connected to the cloud when they leave the factory. The only provisioning needed is to associate the sensors to the point they are measuring, which is done in seconds by selecting the point in an app and tapping the sensor with NFC or scanning a QR code from the phone. Identifier Links enable Industry 4.0 standardized unique identification of every sensor.

Operation and support are handled by Trisense as part of the service. Every sensor reports health metrics and connectivity status to our operations team and is remotely configured and updated to monitor and optimize its performance. Sensors, connectivity, operations, data storage, analysis and dashboards are all delivered as a subscription service, enabling the customer to focus on utilizing the insights provided by CirruSense.

Finally, battery life is crucial to the business case, as replacing sensors is an expensive operation. By turning every stone in electronics, firmware, communication protocol and cloud platform, the sensors have a battery life of 15 years with one measurement per hour and one transmission per day.

From raw data to insights

With dense coverage of sensors continuously monitoring the conditions of piping systems comes vast amounts of data that must be securely and robustly transferred and stored. However, measurements by themselves do not yield much value.

First, the data must be structured in such a way that they can be processed and refined. As sensors might be moved or replaced, a system that enables mapping between different sensors and logical measurement points to different time periods is needed. This allows focusing on analyzing continuous data sets without concerning the history of physical sensors. Furthermore, measurement points are organized in a structure that reflects the customer's assets and their properties, making it possible to efficiently analyze data and build insights specific to the CUI domain.

Next, measurements must be normalized and sanitized to ensure the quality of data. As an example based on field experience, we know that sensors in the same plant are differently affected by sun radiation, causing undesired variations in measurements irrelevant to the wetting state. By compensating for this, the interesting variations in the data appear. This is performed in the cloud together with other calibrations and corrections.

States of the monitored assets are then classified, so multiple measurements of continuous ranges are distilled down to classifications indicating whether the monitored point is dry, humid or wet for every day. Furthermore, the classifications are aggregated over time and over multiple measurement points, to create top-level reporting that ultimately helps optimize maintenance plans.

Finally, as the value addition described above happens in the cloud, it is possible to improve results as refinements are made to algorithms and new features are developed.

From insights to action

As measurements are collected, organized, analyzed and aggregated, top-level reports reveal the risk levels of different monitored assets. The reports can be viewed in interactive dashboards or presented in PDFs. In any case, the main purpose of the insights is to enable optimization of maintenance plans. By knowing the degree and location of wetting for every asset, it is now possible to spend the maintenance resources where it matters most.

Another valuable possibility is to perform spot repairs when water intrusion occurs. The event detection system of CirruSense notifies the operators by e-mail, allowing immediate action before the water degrades the coating and eventually leads to corrosion. This helps maintain the integrity of the assets and is especially helpful on new or newly refurbished assets.

Finally, continuous monitoring documents the integrity of the assets, as well as the effectiveness of repairs performed.

Future possibilities

Data collected over several years can increase our understanding of how different factors affect corrosion of the pipes. By utilizing tools such as machine learning together with feedback from inspection results, continuous wetting data, as well as ongoing research into other factors such as coating degradation mechanisms and effects of temperature cycling, we will be able to provide highly precise predictions of CUI risk.

147b Adacta: A Research Data Management Solution with a Focus on Traceability

Hendrik Goßler, Johannes Riedel, Eric Daymo, Olaf Deutschmann

Industry 4.0 is in part focused on the digitization of equipment and processes, including the development of digital twins that accurately capture key characteristics of their physical counterparts. In catalyst testing (as is the case in other experimental fields), the test stand configuration contains many important details that are necessary to save in order to fully understand the data that are generated. A digital twin, which contains all the data and metadata necessary to fully represent a physical object or system, is an ideal Research Data Management (RDM) tool. This is because a digital twin can accurately reconstruct exactly what was being tested and which devices were used to collect data. Incorporating digital twins into a catalyst testing RDM system ensures that the data are stored alongside the appropriate metadata so that the experimental results are understandable far into the future. Recognizing the benefits of digital twins for enhancing data reuse, omegadot developed Adacta [1], a novel RDM system designed to create a digital twin of an experimental system, with features specifically incorporated to handle the unique complexities associated with catalyst testing.

One such complexity is the fact that catalyst test stands - either the equipment and/or what is being tested – is often changing. However, keeping a digital twin up to date becomes complicated when the physical twin is constantly evolving. This is often the case with catalyst test stands, which are typically designed to evaluate many different samples. Further complicating the development of digital twins, catalyst test stands are typically reconfigured at various times in order to allow the collection of data under vastly different conditions or even chemistries. As a result, the digital twin of a catalyst test stand today can be vastly different from the digital twin of the same catalyst test stand tomorrow.

Adacta addresses this challenge by organizing data and metadata around the timeline, allowing the exact configuration of a test stand at any given point in time to be saved. In this way, the digital twin remains updated even as new catalysts are tested and the test stand evolves. Equipment changeouts, calibrations, and even new process flow configurations are readily traceable with Adacta. This novel RDM approach thus allows data to be placed in the context of the equipment configuration at the time the results were collected. Moreover, the graphical user interface allows one to locate and visualize data, equipment configurations, and installed test samples in intuitive and flexible ways, such as timeline views, data plots, and annotated process flow diagrams.

Conventional catalysis RDM meanwhile often relies on a "forms-based approach," such as a spreadsheet that contains rigid, predefined, fixed fields for data input and output. While it is technically possible to design a form that can capture the information from the test stand, there are several aspects that make this approach impractical in many situations. For example, it is often difficult to design the form in an unambiguous way. Suppose there is a test stand with multiple temperature sensors installed, and these sensors can be placed in different positions over the life of the system. Labeling a field as "temperature" adds ambiguity to the metadata because it is possible that fields can be interpreted differently by the person entering the data and the person reading (and reusing) the data at a later point in time. While it is certainly technically possible to design the form using detailed descriptions of the fields, considering that catalyst test stands frequently change with time, a significant amount of work will go into maintaining the form such that it reflects the current state of the test stand.

Adacta is highly flexible in that it does not rely on forms to categorize information. Instead, Adacta connects data to devices, which has the advantage of recording information in an intuitive manner that mimics how laboratory data are collected. As an example, a classic RDM approach to record catalyst mass involves storing the sample mass along with the measurement date/time in a specific field on a form, such as a spreadsheet. The Adacta way of storing this information is akin to the actual experimental workflow, whereby the sample is placed on a scale, and the scale records a sample weight at a given date/time. The relationship between the sample, the scale, and the data is made within Adacta during the data import process. By storing the sample weight in a manner that associates the data with both the measurement device and the sample, there is no need for a spreadsheet-like form to save results. This relationship between data and devices also makes Adacta extremely flexible. Unlike form-based approaches to RDM, the data structures within Adacta never become outdated as test stands evolve. As well, even though Adacta was developed with the experimental catalysis community in mind, flexibility in how data are stored makes this RDM approach extendable to any type of experimental system where equipment records data in any format.

Ultimately, as databases grow, organizations will be able to continually mine stored information for new insights, both to troubleshoot and to discover breakthroughs in relations that were previously masked and nearly impossible to find. Especially since it is often not known how experimental results will be used in the future, the chance for data re-use is higher if as much metadata are recorded. Unencumbered by the rigidity of forms that can sometimes miss critical metadata, Adacta enhances the promise of Industry 4.0 in the field of catalyst testing by ensuring digital twins are time accurate, with the exact configuration of test equipment accurately saved alongside the measured data.

References

[1] H. Gossler, J. Riedel, E. Daymo, R. Chacko, S. Angeli, O. Deutschmann, CIT, 2022, 94, 1798.

147c How Adding Machine Learning to Simulation Brings Greater Value for Process Industries

Justin Hodges, Thomas Eppinger, Ravindra Aglave

Machine learning (ML) had surges in popularity every decade before breaking the pattern to yield a permanent adoption of ML into numerous aspects of business, engineering, and society as a whole. Many factors, including but not limited to advances in parallel computing capabilities from GPUs to facilitate the inception of more sophisticated and capable ML algorithms, have supported this AI industrial revolution. Engineers today have the distinct opportunity to utilize these advancements in ML capabilities for our domain area of simulation and physical testing.

For the process industry, the digital tools for design and operation of equipment is ripe for supplementation with machine learning. Design engineers can benefit from the constant stream of innovations in the ML community when productizing new equipment, while operators of the equipment can be enhanced in productivity from on-demand information when these ML models are deployed for their easy access. There are several themes by which machine learning can cause a material impact: augmented user experiences, optimization, better utilization of physical measurement, and time savings via faster insights.

Utilization of a limited number of experiments and complimenting them with simulated data can result in significant time savings and valuable insights into engineering processes. Specifically, how simulation experts, a widely used engineering method, can realize large time savings when employing ML models in their sensitivity and design exploration, which can help reduce time to market. Further, how physical measurement of in-operation equipment can be combined with simulations with as input to (ML powered) executive digital twins (xDTs) to provide the operator with simulation-level detail of performance from limited discrete physical measurement points.

In this study, we perform multiple batches of computational fluid dynamics (CFD) simulations to characterize the performance of a mixing vessel with respect to changes in the geometric design. More significantly, we then explore a series of three independent scenarios whereby unique ML-based ROMs are created and analyzed. First, we conduct an optimization study in a hybrid modality: machine learning models are used when possible, during the optimization study in conjunction with traditional simulation for the design points. Second, we create and leverage ML-based ROMs for adjacent design characterization akin to typical robustness, reliability, and sensitivity analysis. Finally, we create a third ML-based ROM that can impute simulation-level detailed results from only discrete measurements taken while the equipment is in operation.

Our observation is that the impact of simulation-based, detail-dense ROMs, is under-demonstrated in industry and has tremendous value; operators can take typical measurements in their equipment and as such use it as inputs to the ML-based ROMs created. These ML-based ROMs will provide fully 3D results, full of information as if a simulation model was ran, from only a few discrete measurement points (such as a RPM or torque measurement, which are 0D in nature).

These models can be deployed in web-based application to bridge the designer to the operator. It will aid in responding to changes either due to operating conditions or input conditions at a rapid pace while maintaining quality and safety of chemical operations at quicker pace.

178a A Chemical Industry Case Study: Using AI to Create More Productive Teams in the Chemical Manufacturing Industry

Felix Monschein

The development of an effective and efficient search functionality based on AI looked to solve two challenges cited by Bayer:

1. Inefficient access to documented historical knowledge captured by Shiftconnector; and
2. Incomplete entry items where often descriptions of solutions to problems (in the production process) were missing or incompletely described. The development of AI's Smart Search functionality solved both problems. Teams have improved plant performance by quickly finding solutions based on the history of documented tribal knowledge from plant operations teams. By retrieving the most relevant information, it accelerates operations and brings capabilities to help teams recover from disruptions by immediately identifying appropriate fixes. Also, captured knowledge can be managed to support workforce changes.

Eschbach's questionnaires and analysis of usage data show the following advantages of Smart Search:

- Smart Search is well accepted and used several times daily.
- The time users spend searching for information decreased significantly. Instead of searching for several minutes, users get results within seconds.
- User statistics show that users access all historical knowledge (> 8 years) captured in Shiftconnector. Prior to the Smart Search introduction, only records from the last month were accessed.
- Statistics and questionnaires show that now the historical know-how captured in Shiftconnector (solutions to problems in the production process) is effectively used by all users of Shiftconnector, from shop-floor operators to process engineers.

The Shiftconnector Smart Search feature changes the way shop floor operators in the process industry search. The Shiftconnector Smart Search feature was developed using the latest technology and architecture for AI-based language models. This allows the Smart Search feature to understand the context of the search versus a standard keyword search that would need very specific terms to retrieve relevant information. With Machine Learning, the smart search retrieves relevant information with a simple, everyday phrase like "why is this product brown instead of gray?"

Search engines in today's market like Google are smart but do not think like someone in the process industry as they lack industry-specific or even plant-specific terminology.

Shiftconnector's Smart Search is superior in this aspect as it was developed with a lexicon of the operators in the process industry. With nearly 20 years in the industry and 70,000 global users, the eschbach team understands not only the specific use cases for a search engine but also the specific terminology that is used every day on the plant floor. Consequently, Smart Search covers plant-specific terminology, such as searching for technical labels which are only used at a certain production line. For example, if in a plant the label "CP001" is used for a centrifugal pump at a reactor inlet, Smart Search understands this and will retrieve relevant information linked to "CP001" when an operator is searching for "centrifugal inlet pump".

178b Generative AI for Process Systems Engineering: Potential Applications and Pitfalls

John Hedengren, Junho Park

This presentation is an introduction on Generative AI on the field of Process Systems Engineering (PSE). At its core, PSE revolves around the mathematical modeling, optimization, and simulation of chemical and physical processes. With Generative AI, particularly advancements like Large Language Models (LLMs), offer a new avenue to complement and enhance development of these traditional methodologies. In related work areas, the productivity of consultants improved by 25% and the quality of the work increased by 40%. This presentation provides an overview of new Generative AI tools, with applications for workforce development and time-series modeling with Tranformers. AI's capability in conceptual process control design will also be discussed, helping process control engineers develop new systems or revamp existing ones. However, as with any new technology, challenges exist. There is a distinct jagged edge of application beyond which the Generative AI provides incorrect information. There are additional issues of over-reliance, the explainability of the AI outputs, and the repeatability of AI-backed processes. General purpose tools can be adapted to specific company applications with transfer learning to reduce the required data and training requirements.

4a Sustainability, Digitalization, and the Path to Zero

Billy Bardin

Today's market and stakeholder demands require institutions to produce materials more sustainably, with demonstrated progress towards reductions in greenhouse gas emissions, more efficient water use, and improved ability to recycle products at the end-of-life. Dow has embarked on a "decarbonize and grow" strategy that will drive towards reducing scope 1 and 2 emissions by 5 million metric tons between 2020-2030, while advancing on its longer term, 2050 objectives. The intersection of digital capabilities with sustainability strategy is a necessity. The reporting, tracking, and transacting of carbon emissions require improved data connectivity and transparency enabled by a digital thread through the value chain starting with suppliers and flowing through to the customer. Improved monitoring of assets and optimization of processes help to validate emissions data while enhancing operational performance. Examples of current and planned programs supporting sustainability ambitions will be highlighted.

148b Transformer-Based Hybrid Modeling and Control of Evolving, Nonlinear Processes

Joseph Kwon

Traditionally, the dynamic modeling of chemical processes has relied on first-principles models grounded in fundamental physics and chemistry laws. These models, primarily formulated through differential equations with constant parameters, enable the calculation of control actions optimizing process operations, taking both process and actuator limitations into account. However, the ever-evolving and nonlinear nature of chemical processes frequently calls for models with time-varying parameters. Motivated by these challenges, we have developed hybrid models that integrate system-agnostic first-principles dynamics with system-specific data-driven, time-varying parameters. Our hybrid modeling framework incorporates a recent innovation: attention-based time-series transformers (TSTs) coupled with positional encoding. This marks a pioneering venture into applying the transformer algorithm – a cornerstone in ChatGPT's triumph - to nonlinear, time-varying processes. By analyzing data across both current and preceding time steps, the TST captures both immediate and historical changes in process states, granting a contextual insight on process dynamics, mirroring ChatGPT's textual context understanding. This TST-based hybrid model identifies correlations between process parameters and state variables. Its versatility is evident as it adapts to a spectrum of models - from density function theory to computational fluid dynamics - and scales, spanning from laboratory to extensive industrial environments. We will present applications of this hybrid modeling and control architecture, showcasing its utility from labs to industrial processes, made possible through partnerships with leading chemical process enterprises.

BIO: Joseph S. Kwon is an Associate Professor in the Department of Chemical Engineering at Texas A&M University (TAMU), where he holds the esteemed Kenneth R. Hall Career Development Professorship. He earned his B.S. in Mathematics and Chemical Engineering from the University of Minnesota, Twin Cities. Subsequently, he obtained his M.S. in Electrical Engineering from the University of Pennsylvania and a Ph.D. in Chemical Engineering from University of California, Los Angeles. Dr. Kwon specializes in the development of computational models and frameworks within the field of Process Systems Engineering (PSE). His research focus encompasses multiscale modeling, hybrid modeling, model predictive control, and optimization for diverse chemical and biological process systems. He has published over 120 papers in top-tier journals, and delivered numerous invited talks at prominent chemical process entities such as ExxonMobil and Halliburton, universities, and conferences. To date, he has secured over $4 million from agencies like the NSF, DOE, as well as from private industry. He has received multiple awards, including the Hanwha Non-tenured Faculty Award, KSEA Young Investigator Award, KIChE President Young Investigator Award, and James M. Lee Memorial Award. Within TAMU, he has received the Engineering Genesis Award, TEES Young Faculty Fellow Award, and Distinguished Teaching Award from the Department of Chemical Engineering. Additionally, he has secured best paper awards from both the IET Systems Biology Journal and the Mathematics Journal.

166a Hybrid Modeling for Critical Velocity Predictions for Solid Transport

Su Meyra Tatar, Yushi Deng, Haijing Gao, Shuxing Cheng, Simona Duplat, Selen Cremaschi

Accurate critical velocity predictions in multiphase flows are crucial for solid transport where the critical velocity is the minimum velocity that the solids are transported without any deposition in wellbores/pipelines (Dabirian et al., 2016). The hybrid modeling approach, which is a combination of the first principal model and data-driven model, is one of the methods that can be used in predicting the critical velocity. This work adopts the parallel structure of a hybrid modeling approach for critical velocity and its uncertainty predictions (Sansana et al., 2021) (Deng et al., 2022) where the Oroskar and Turian (Oroskar & Turian, 1980), Mantz (Mantz, 1977) and Tulsa Models (Najmi, 2015) are used as the semi-mechanistic models and the Gaussian Process Modeling (GPM) (Rasmussen & Williams, 2006) is used as the data-driven model. Fig 1 represents the parallel structure of the hybrid modeling approach where and ^ represents the predictions made by the semi-mechanistic model and hybrid model respectively. Also, the δ and δ̂ represents the discrepancy and discrepancy estimates by the data-driven model where the discrepancy is defined as the difference between the critical velocity measurements (), and critical velocity predictions by the semi-mechanistic model (). Additionally, terms represent the input data set which includes solid density and solid concentrations, fluid viscosity and densities, and hydraulic and particle diameters. The GPy is used as the Gaussian Process (GP) framework to train the GPM model for the estimation of model discrepancy where the method of k-fold cross-validation is used for training the model with four folds. This work discusses the hybrid modeling framework and analyzes the findings from three semi-mechanistic models and three hybrid models. Also, it compares the performance of these resulting models for different ranges of critical velocity measurements

166b Network Optimization to Enable Transition to a Sustainable Chemical Industry

Scott Bury

This talk presents an exploration into how Operations Research (OR) contributes to decarbonization and sustainability efforts, particularly highlighting the innovative approaches undertaken by the MiLOS team. By leveraging advanced OR methodologies, the group supports achieving Dow's ambitious decarbonization and sustainability targets in partnership with functions and businesses across Dow. The talk will delve into case studies demonstrating the application of optimization, simulation, and decision analysis to reduce carbon footprints, enhance energy efficiency, and promote sustainable practices across various operations.

166c Integrating Physics and Machine Learning: A Hybrid Modeling Approach for Electrochemical Separation

Teslim Olayiwola, Revati Kumar, Christopher Arges, Jose Romagnoli

Electrochemical separation plays a pivotal role in a variety of industries, ranging from water treatment to oil refineries. Despite its efficiency, the challenge lies in the design and synthesis of targeted processes. Mathematical models are crucial in this context, as they facilitate the development of new technologies by enabling simulation, sensitivity analysis, techno-economic evaluations, and optimization. These models are instrumental in providing insightful information while curtailing the time and resources necessary for devising new designs. The integration of data-driven and knowledge-based modeling approaches can significantly hasten the innovation of new technologies.

In our study, we present a novel approach where physics-based models generate synthetic data imbued with specific material properties. This data is then used to train a machine learning-based surrogate model on which Transfer Learning are applied to fine-tune the model's performance, aligning it more closely with experimental measurements. Our modeling approach is tailored for systematic screening of material properties and selecting optimal parameters for guiding electrochemical separations. The effectiveness of this methodology is exemplified through its application in modeling two widely used water desalination technologies: electrodialysis and electrodeionization.

56a Practical Application of a Mixture-Amount Design of Experiment to Optimize a Complex Product Formulation.

Brian Clark

This presentation will discuss a case study that delves into the practical application of a mixture design of experiments to optimize a complex product formulation. Mixture experiments are a valuable tool in industries where products are composed of multiple components, each contributing to the overall quality and performance.

The methods and techniques employed in the mixture design of experiments are discussed, including the choice of factors and levels. The study demonstrates the data collection and the statistical analysis used to generate models and optimize the product's formulation.

The results section presents the findings of the study, emphasizing the significant improvements achieved in the product's quality and cost-effectiveness through the optimized formulation. Key insights gained from the experiment, such as identifying the most influential components and their optimal proportions, are highlighted.

In conclusion, this case study illustrates the effectiveness of a mixture design of experiments in product development and optimization. It serves as a reference for professionals and researchers in industries where product formulations involve complex mixtures and the need for efficient, data-driven solutions is paramount.

56b Jmp-Fastman: All-in-One Tool for Data Exploration and Fault Detection Analysis

Kyle Territo, Jose A. Romagnoli

In the era of big data, effective data exploration is paramount. An essential requirement is a versatile tool capable of seamlessly navigating vast datasets to extract critical insights, unleashing the full potential of data understanding. By combining dimensionality reduction, clustering, and machine learning techniques, we usher in a new tool for data comprehension.

JMP-Fastman emerges as a dynamic add-on to the existing JMP software by SAS. Leveraging the robust JMP platform, we harness its extensive array of features, particularly its powerful visualization capabilities. Moreover, the visualizations created maintain seamless connectivity with the underlying data, allowing users to effortlessly interact between plots and data tables.

The primary objective of Fastman-JMP is to provide users with a fresh perspective on big data, making it more comprehensible and actionable. Following data preprocessing, dimensionality reduction, and clustering algorithms come into play. These transformations unveil underlying trends and potential data anomalies or faults. We gain insights into how processes evolve over time, with the ability to pinpoint the variables most responsible for deviations or faults. These reduced projections can be visualized in 2D and 3D, facilitating comparisons between different dimensionality reduction and clustering algorithm outputs. Additionally, the tool offers valuable features such as sensitivity analysis, hyperparameter optimization, and self-organizing maps.

After the culmination of the data exploration phase, the tool empowers users with the capability to fashion predictive models, thereby expanding its utility beyond mere analysis. Fastman-JMP offers an array of robust features for crafting both classification and regression models, capitalizing on the rich insights extracted during the initial data exploration process. This multifaceted functionality proves especially invaluable in industrial settings, where it finds practical application in tasks such as the development of soft sensors tailored for monitoring operational data. In essence, Fastman-JMP not only illuminates data intricacies but also propels users towards data-driven decision-making and predictive excellence, enhancing its role as an indispensable asset in the world of data analytics.

In this presentation, we will delve into real-world use cases, demonstrating the versatile applications of Fastman-JMP across a wide spectrum of fields. Fastman-JMP is poised to redefine the landscape of data exploration and fault detection, empowering users to unlock hidden insights within their data.

56c Real-Time Design Modeling for Control Valve Size and Type Using Machine Learning and Big Data Streaming System

AReum Lee

In the chemical processing facility, the control valve assumes a crucial function in the regulation of fluid flow and pressure. Also, the selection of an appropriate valve type is intricately linked to the operational efficiency and safety of the process. The selection of valve types relies on the expertise and experience of specialists, necessitating a considerable investment of time and financial resources. To solve the problem, we proposed a method for predicting valve size and type based on machine learning approaches. To analyze the predictive performance of the model according to the phase of the fluid, we extracted the dataset by dividing the data of three fluid phases: liquid, gas, and steam. Additionally, the proposed method was evaluated using a F1 score and a receiver operating characteristic (ROC) curve as assessment metrics.

Moreover, in this study, process data is transmitted in real-time to a cloud server through Kafka, a data streaming system. For the prediction of control valve size and types, the machine learning analysis model is executed on a cloud server, and the prediction results are transmitted in real-time to the process and control engineering platforms through Kafka.

Therefore, the application of the proposed method in the design process, it is anticipated a reduction in the time and cost associated with determining control valve size and type.

56d AI-Enabled Dynamic Risk Register: A Novel Approach to Risk Management

Bao Lin Tan

Process safety accidents and incidents can have devastating consequences, both in terms of human life and financial costs. Traditional process safety risk management solutions are often limited in their ability to prevent these accidents. They may be based on outdated data or rely on manual processes that are prone to error.

The AI-Enabled Dynamic Risk Register is a novel approach to process safety risk management that overcomes these limitations. It uses artificial intelligence to analyze historical and real-time data from operational, process parameters, maintenance, and reliability data to predict future risks, recommend remediation actions, and identify the root causes of risks.

The AI-Enabled Dynamic Risk Register uses artificial intelligence to analyze historical and real-time data to predict future risks, recommend remediation actions, and identify the root causes of risks. The system is divided into three main components:

1. **Foundational models:** By using hybrid models that combine computer vision, time-series forecasting, and physics-based modeling to diagnose and predict the condition of heat exchangers, furnaces, and pumps. These models are more accurate and robust than individual models, and they can provide more insights into the condition of the equipment.
2. **Prescriptive models:** LSTM prescriptive models analyze the predicted results from foundational models to prioritize risks and recommend remediation actions. LSTM prescriptive models are particularly well-suited for this task because they can learn long-term dependencies in the data, which is important for predicting equipment failures and recommending remediation actions.
3. **Dashboard:** An interactive dashboard using Power BI that displays the overall risk, top threats, and recommendations to reduce risk.

The AI-Enabled Dynamic Risk Register has been shown to be effective in reducing accidents and incidents by up to 20%, maintenance costs by USD 1 million, and loss of revenue by USD 4 million. For example, one company that implemented the AI-Enabled Dynamic Risk Register was able to prevent a major accident that could have cost millions of dollars in damage and lost revenue.

The AI-Enabled Dynamic Risk Register is a valuable tool for any company that wants to improve its process safety performance. It is a powerful example of how digitalization can be used to improve safety in industrial plants.

180a Iadys: Innovate, Clean-up, Preserve

Cyril Castello, Andreas Kampouris, Damien Boursiquot

The IADYS Jellyfish Bot stands as a groundbreaking depollution solution meticulously crafted for the diverse landscapes of maritime, river, and lake harbors, along with extensive waterways. This compact and robust robotic marvel is designed to autonomously or remotely address the pressing issues of floating waste and oil. The Jellyfish Bot effectively curtails the dispersion of pollutants, safeguarding rivers, seas, and oceans from environmental harm.

Initially conceived for pollution control in ports, the Jellyfish Bot's ingenuity lies in its small yet powerful design. Its diminutive stature allows it to navigate and treat the most inaccessible spaces, deftly maneuvering between boats and beneath pontoons. This unique attribute ensures a comprehensive and targeted approach to pollution management, reaching areas that would otherwise be challenging to address.

Autonomy is at the core of the Jellyfish Bot's operations. It seamlessly undertakes its missions, skillfully avoiding surface and underwater obstacles. This autonomy not only enhances its efficiency but also minimizes the need for human intervention, making it a reliable and resilient solution for depollution efforts.

Functioning as a true 'Swiss Army knife' of the port, the Jellyfish Bot excels in multifaceted roles. Beyond its primary function of collecting solid waste and oil, it extends its capabilities to perform bathymetric mapping. This dual functionality positions the Jellyfish Bot as an indispensable tool for comprehensive environmental monitoring and management within port ecosystems.

In essence, the IADYS Jellyfish Bot transcends conventional depollution methods. Its adaptability, autonomy, and versatility make it a pivotal player in the ongoing efforts to protect and preserve aquatic environments, offering a sustainable solution to the complex challenges posed by pollution in diverse water bodies.

Industry 4.0 Applications to Three Engineering Operations – A Learning Experience for Senior Engineering Students

Richmond Boone*, Riley Davis, Jared Starnes, and Bahman Ghorashi

Chemical Engineering Department
Tennessee Tech University

Abstract
Industry 4.0 is characterized by technologies that connect the physical and digital space. It could be construed as an evolution of Industry 3.0 which brought us the digital revolution. It will encompass the evolution of technologies, such as: robotics, nanotechnology, artificial intelligence, internet of things (IoT) as well as the industrial internet of things (IIoT), bio-molecular science, blockchain technology, additive manufacturing, autonomous vehicles, and fifth generation wireless technologies. These technologies will significantly speed up the rate of change and will drastically alter the way companies will produce, market, and distribute goods and services and develop life-long relationships with their customers.

The fourth Industrial Revolution can be defined as the connection between the physical world and digital spaces. It allows machines to communicate with humans and with one another to accomplish tasks that were not previously possible. For instance, digital twinning allows one to view and manipulate a real-time simulator, populated by real data, and big data can store and analyze monumental amounts of data. Industry 4.0 technologies have the potential to completely change how industries operate, making them both safer and more efficient. It could even lead to "lights out" facilities, where no human intervention might be necessary to run a process. There are, however, technological, social, and ethical challenges that need to be addressed.

Chemical companies are not oblivious to the transformative changes that Industry 4.0 will bring. Many are already changing or preparing to change their mode of operations to incorporate strategic goals that would make them more competitive. Workforce preparation is a part and parcel of such a major transformation. Hence, it is prudent for technological colleges and universities to offer interdisciplinary courses, minors, or concentrations that in some way would bridge the gap among various Industry 4.0 disciplines, thereby allowing students to broaden their perspectives and explore possibilities beyond their narrower academic field.

Today's information workers need to be multi-taskers, fully capable of working with colleagues, suppliers, and customers with vastly different educational backgrounds. Hence, in addition to

* Mechanical Engineering Department

their in-depth knowledge in their own areas of expertise, they also need to have a broader understanding of other relevant academic disciplines.

This paper specifically describes three selected chemical engineering operations that were significantly transformed by a group of students who incorporated selected Industry 4.0 technologies. The three operations included injection molding, wastewater treatment, and synthesis of iron nanoparticles. Student engagement was in form of course projects conducted by individual students as they evaluated selected operations and explored ways to transform those operations into "lights-out-manufacturing" with little or no human intervention.

The objective was to make the operations completely automated using I-4.0 technologies. Technologies that were considered included: IIoT, ML, blockchain, AI, AR, and VR. Additionally, upstream, and downstream operations were considered with particular attention to industrial symbiosis possibilities.

Introduction

Over the past year, the senior author, Bahman Ghorashi, developed a new course covering the key technologies that collectively form the basis for Industry 4.0. The course was constructed to explore the areas of overlap among various engineering disciplines and to build on common topics familiar to all engineering students. This was done by stimulating and supporting students with different engineering backgrounds to learn the interrelationships and commonalities among the constitutive disciplines that shape the evolvement of Industry 4.0. The premise was that the critical role of educational institutions in preparing the future workforce for Industry 4.0 should be addressed through courses and concentrations that would facilitate the closing of the existing gaps.

This paper addresses the development of a new course covering the above-mentioned areas and delves into summarizing three chemical engineering operations that were significantly transformed by a group of students who incorporated selected Industry 4.0 technologies. The course is in part based on the senior author's experience of teaching a course in Agile Manufacturing/Agile Organizations that has been received enthusiastically by students and has been taught at three different universities in the United States and Germany.

This new course covered various aspects of Industry 4.0 , engaged students in project work as well as traditional class lectures. Agile manufacturing concepts were covered together with case studies that examined actual initiatives as organizations began the transformation process.

Students who successfully completed the course acquired the knowledge of how companies strategically plan to transform themselves and seamlessly incorporate and utilize the next generation technologies and agile principles in their modes of operation. Such a knowledge would aid the engineering and businesses graduates to better understand and appreciate the decision-making process as they begin their professional careers. Additionally, their readiness to contribute from day-one will save their employers precious time that otherwise would have to be spent on training and retraining.

Topics included:

- Industry 1.0, 2.0, and 3.0
- Industry 4.0 Definition and Overview
- Industry 4.0 Technologies
- Challenges Facing Industry 4.0
- Quality Control
- Supply Chain Management
- IoT - The Internet of Things
- IIoT – Industrial Internet of Things
- Big Data
- Automation & Software Technologies
- Virtual Reality VR, Augmented Reality AR, Artificial Intelligence AI
- Overview of Mass Manufacturing, Lean Manufacturing and Agile Manufacturing
- Four Pillars of Agility
- Characteristics of Agile Organizations
- Agile Principles
- Flexible Production
- Digital Factory
- Digital Manufacturing Operations
- Design Modularity in Smart Factories
- Smart Sensors, Sensor to Cloud Connection
- Edge Computing
- Digital Twins
- Machine Learning
- Data Warehouse
- Flexible Manufacturing Systems
- Additive Manufacturing
- Machine to Machine (M2M) Communication
- Cryptography and Blockchain Technology
- Flexible Mass Production
- Mass Customization

Student Engagement

Student engagement was in form of course projects conducted by individual students as they evaluated individual engineering operations and explored ways to transform those operations into "lights-out-manufacturing" operations with little or no

human intervention. Four different engineering operations were selected by students. The objective was to make the operations completely automated (lights-out-manufacturing) with little or no human intervention. Technologies that were considered included: IIoT, ML, blockchain, AI, AR, and VR. In their analysis, they also included upstream and downstream operations and considered industrial symbiosis possibilities. This paper provides a summary of the students' findings and from those draws conclusions related to students' learning experience.

First Operation - Synthesis of Iron Nanoparticles

Iron nanoparticles possess special characteristics applicable in diverse fields, including targeted drug delivery, cancer treatment, and nano-coatings for materials. However, achieving consistent production of these nanoparticles is challenging. A reliable method for controlling their growth involves a thermal decomposition process. Currently, a detailed procedure utilizing a Schlenk line is employed for particle production. This study aims to discover an improved, large-scale production method for accurate results, incorporating Industry 4.0 principles. Artificial Intelligence plays a crucial role in communication within the supply chains, connecting various sensors throughout the process via the Internet of Things. Automation and advanced robotics are expected to reduce costs and enhance efficiency. The implementation of this process considers ethical considerations, predictive maintenance, digital twins, and quality inspection of the particles to produce high-quality nanoparticles in mass quantities.

Iron nanoparticles present promising solutions to challenges in biomedical engineering, offering applications in magnetic particle imaging, hyperthermia treatment, and MRI contrast enhancement[1]. Beyond biomedical engineering, iron nanoparticles can be used in various fields, as highlighted in the National Institutes of Health (NIH) article, "Synthesis, Characterization, Applications, and Challenges of Iron Oxide Nanoparticles"(Ali et al., 2016). This research identifies applications in biomedical, health care, agriculture, food, environmental remediation, energy, defense, aerospace, construction, automotive, textiles, and electronics.[2]

Given the considerable potential of iron nanoparticles, it becomes imperative to address the need for precise and accurate large-scale production. The integration of Industry 4.0 technologies and advanced automation offers a pathway for achieving consistent large-scale production and nationwide distribution. Effectively implementing Industry 4.0 technologies in the synthesis of iron nanoparticles requires a comprehensive consideration of the entire process, encompassing the upstream supply chain to the downstream customers.

The production of iron nanoparticles involves three main stages. To begin, the creation of a surfactant involves using an HDA-HCl mixture. Secondly, the surfactants and precursor are combined in a round bottom flask, and a Schlenk line is utilized for particle synthesis. Lastly, the particles undergo washing, storage, and transportation. Due to the complexity of all three stages, this report focuses on the second stage— the synthesis of iron nanoparticles. It is assumed, for this purpose, that iron nanoparticles can be produced at a large scale with all inconsistencies resolved.

---

[1] Caroline Billings et al., "Magnetic Particle Imaging: Current and Future Applications, Magnetic Nanoparticle Synthesis Methods and Safety Measures," *International Journal of Molecular Sciences* 22, no. 14 (July 17, 2021): 7651, https://doi.org/10.3390/ijms22147651.

[2] Attarad Ali et al., "Synthesis, Characterization, Applications, and Challenges of Iron Oxide Nanoparticles," *Nanotechnology, Science and Applications* Volume 9 (August 1, 2016): 49–67, https://doi.org/10.2147/nsa.s99986.

To successfully integrate Industry 4.0 principles into the synthesis of iron nanoparticles, a collaborative effort among various companies is essential, one where technologies seamlessly connect through the Internet of Things (IoT). From artificial intelligence managing inventory to wireless sensors interfacing with the IoT and advanced automation, each component plays a crucial role in realizing large-scale, lights-out manufacturing processes for iron nanoparticles. Although much of this technology is in its early stages, some companies have already succeeded in developing the equipment required for lights-out manufacturing.

Prior to introducing Industry 4.0 into the synthesis of iron nanoparticles, the current method for creating them poses challenges. This includes the time-consuming and intricate nature of tasks like preparing the surfactant (HDA-HCl) a day in advance, requiring overnight drying in a vacuum oven. Manual weighing and dispensing of chemicals in a flask can be inaccurate for precise measurements. Tasks like retrieving liquid nitrogen are both time-consuming and potentially hazardous if not handled carefully. The process involves manual start-up, purging, chemical retrieval, and changing gas tanks.

Upon upscaling and incorporating Industry 4.0 technologies, the process undergoes a transformation, depicted in Figure 1, featuring larger bulk chemicals, automatic dispensers, robotics, an automated Schlenk line, and an air separator unit for continuous argon gas and liquid nitrogen supply.

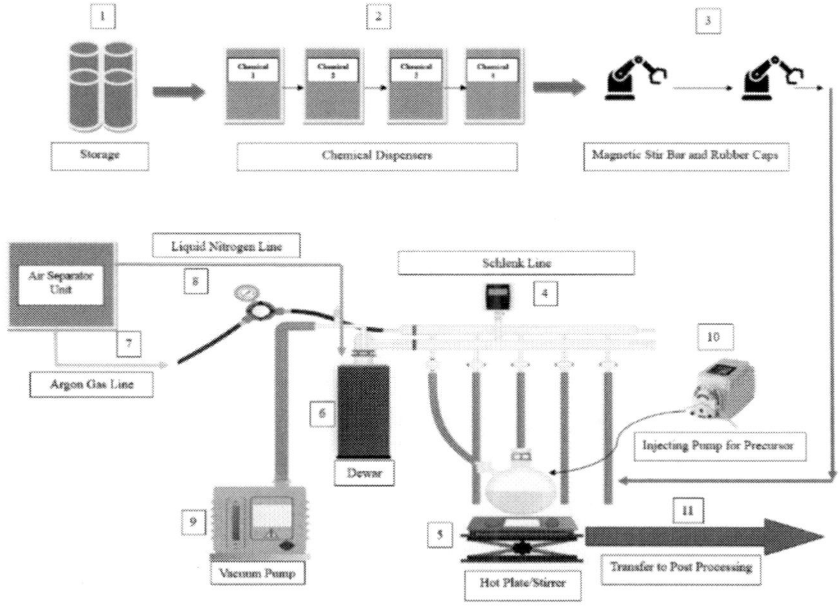

Figure 1 - Large-Scale Iron Nanoparticle Production

In the era of Industry 4.0, the synthesis of iron nanoparticles can undergo a revolutionary transformation, marked by the seamless integration of cutting-edge technologies. This comprehensive overhaul can start with the optimization of the upstream supply chain, introducing efficiency and automation at every stage. To initiate this streamlined process, bulk metal drums carrying crucial chemicals, such as oleylamine, oleic acid, octadecene, and iron pentacarbonyl, can be sourced from chemical suppliers. The utilization of RFID tags ensures efficient tracking and management of inventory, with artificial intelligence inventory management software communicating seamlessly with upstream suppliers. This dynamic interaction not only expedites the acknowledgment of chemical arrivals but also establishes a robust foundation for the automated shipping, handling, and storage process.

A pivotal component of this automated system is the deployment of an autonomous forklift. Such forklifts are equipped with built-in artificial intelligence, ensuring a smooth and efficient transfer of chemicals to the designated storage areas. The synergy of RFID tags, inventory management software[3], and autonomous forklifts[4] provides a fully automated operation for material handling and chemical processing.

Transitioning to the operational requirements of the Schlenk line, the demand for argon gas and liquid nitrogen is crucial. Two potential supply methods can address this situation. The first involves the shipment of large containers to the facility, necessitating manual intervention for tank installation and replacement. Alternatively, the second method employs an on-site air separator unit (ASU)[5]. Advanced PLCs enable the automation of normal operating procedures, eliminating the need for constant supervision. Moreover, these ASUs can be remotely monitored through a cloud platform, ensuring a seamless and secure supply of oxygen, nitrogen, and argon.

With the foundational elements in place, the synthesis of iron nanoparticles commences with a loading process. A three-necked round bottom flask is secured into an aluminum reaction block holder, facilitated by robots. These robots[6], equipped with soft grippers delicately transfer the flask to prevent any damage.

The subsequent dispensing of four surfactants at a designated station occurs through the implementation of automated solutions. Utilizing peristaltic pumps for precision, a digital twin of the dispensers will be instrumental in monitoring performance and collecting essential data. This collaborative integration of technology ensures not only accuracy in dispensing but also provides real-time insights into the usage and type of chemicals.

Continuing along the synthesis line, a sequence of automated steps follows. Robots, equipped with sensors, undertake the insertion of a magnetic stir bar and a rubber stopper into the flask. The entire dispensing stage can be closely monitored to identify any anomalies, ensuring a flawless transition to the subsequent phases of the synthesis process.

---

[3] Cin7, "Inventory Solutions for Product Sellers - CIN7," September 20, 2023, http://www.cin7.com/.
[4] "CYNGN | AI-Powered Autonomous Technology for Industrial Fleets," n.d., http://www.cyngn.com/.
[5] "SIAD Macchine Impianti Air Separation Units | SIAD MI | SIAD Macchine Impianti," n.d., http://www.siadmi.com/air-separation-units.
[6] "Yaskawa Motoman Robotics," Yaskawa Motoman Robotics, n.d., http://www.motoman.com/en-us.

Upon completion of the dispensing stage, the flask progresses to a heating and stirring plate, automated through PLCs. The subsequent connection to the Schlenk line involves spring-loaded U-clamps, securing the flask in place. Temperature sensors can be strategically inserted to monitor the reaction temperature.

A critical facet of the synthesis process involves a diverse array of sensors deployed in the Schlenk line, argon gas, and liquid nitrogen lines. Pressure, vacuum, and temperature sensors play a pivotal role in maintaining optimal conditions throughout the process. Wireless sensors[7] are important in facilitating seamless integration with the Internet of Things (IoT).

As the system reaches connectivity culmination, the stainless-steel Dewar is automatically filled with liquid nitrogen. An automatic Dewar filling station[8] eliminates the need for manual intervention, streamlining operations significantly. Fluid level sensors ensure precise monitoring of waste fluid in the Dewar, triggering a drain valve when required.

The startup of the vacuum pump, an integral component of the synthesis process, sees the application of an oil-lubricated rotary vane system with a PLC controller[9] and provides automation for both startup and shutdown procedures. The argon gas line is further enhanced with the addition of a digital valve controller, regulating gas flow within the Schlenk line. Sensors, including pressure sensors and Pirani gauges, further optimize the efficiency of the inlet valves and the Schlenk line system.

With the system in operation, heating and stirring are applied through automated hot plates. Upon reaching the desired reaction temperature, a robotic arm inserts a syringe needle connected to a reservoir of precursor, injecting the chemical into the flask at a controlled rate. This precise and automated chemical injection is evidence of the unity of advanced technologies in the synthesis process.

Post-reaction, robots disconnect the flask from the Schlenk line, removing the thermocouple and rubber stopper. The flask is then transported to the post-processing station, while the gas flow, vacuum pump, and heating plate automatically transition to idle mode, awaiting the next batch. This entire process is orchestrated to ensure efficiency, precision, and a seamless workflow.

The abundance of data generated throughout these intricate processes is a valuable resource for optimizing and monitoring the performance of the synthesis system. Companies equipped with

---

[7] Scott Dalgleish, "Phase IV Engineering | Wireless Sensor Solutions for the Industrial IoT," Phase IV Engineering Inc., December 22, 2022, http://www.phaseivengr.com/.

[8] Chrisnichols, "Automatic Dewar Filling Station," Technifab Products, Inc., June 7, 2023, http://www.technifab.com/filling-station/.

[9] Q-Vac Priming Systems, "Simplex Oil-Lubricated Rotary Vane Priming Systems - Q-VAC Priming Systems," Q-VAC Priming Systems, September 21, 2018, https://vacuumprimingsystems.com/product-center/simplex-priming-systems/simplex-oil-lubricated-rotary-vane-priming-systems/.

big data software[10], play a crucial role in analyzing and interpreting this data. The insights derived from this analysis contribute to refining the entire synthesis process, ensuring continuous improvement and adherence to stringent quality standards. Figure 2 shows the various Industry 4.0 Technologies implemented throughout the process.

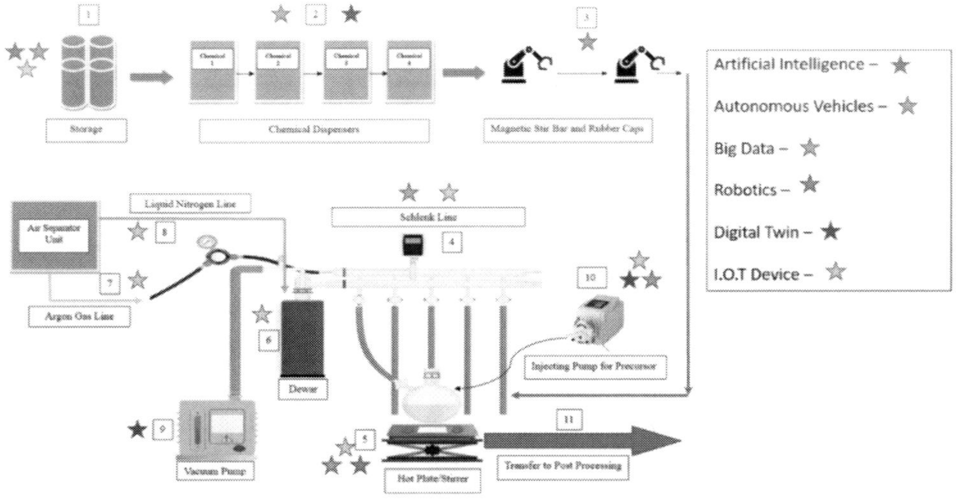

Figure 2: Industry 4.0 Technologies implemented throughout the process.

The introduction of Industry 4.0 technologies significantly impacts the financial aspects of iron nanoparticle synthesis. By fully automating the Schlenk line setup, the need for manual labor is eliminated, reducing costs associated with hiring and managing lab technicians. Artificial intelligence and autonomous vehicles in shipping and handling further contribute to cost savings by minimizing the dependence on human labor. The application of AI in market analysis ensures efficient material storage, mitigating the risk of overstocking and optimizing resource allocation. The incorporation of an on-site air separator unit not only eliminates the continuous expenditure on argon gas and liquid nitrogen tanks but also presents a potential revenue stream through the sale of excess gases or liquids.

The adoption of Industry 4.0 requires a comprehensive approach to risk assessment and maintenance. Digital twins play a pivotal role in monitoring the efficiency of chemical dispensers and pumps. Any drop in efficiency triggers a human intervention for prompt maintenance. Also, resonance ultrasonic vibration sensors[11] can be used in the Schlenk Manifold to detect cracks in glassware, ensuring timely repairs and resumption of operations. Autonomous vehicles equipped

[10] "Turn Data into Insights with Oracle Big Data," n.d., http://www.oracle.com/big-data/.

[11] Ultrasonic Technologies, "Glass Crack and Flaw Detection | Ultrasonic Technologies," n.d., https://www.ultrasonictech.com/glass.

with spill-detection capabilities contribute to a proactive approach to addressing chemical spills during transportation. To mitigate potential hazards, alarm systems are essential to alert personnel in the event of explosions, given the combustible nature of certain chemicals.

Ensuring the quality of iron nanoparticles demands advanced techniques due to their nanoscale nature. Human operators or advanced robotics are essential for extracting particles from the production line for analysis. Quality checks include assessing the magnetism using a Gauss meter, measuring particle size through dynamic light scattering, and employing microscopy techniques such as SEM, TEM, or AFM[12] to scrutinize chemical composition. These meticulous quality control measures are integral to guaranteeing the consistency and reliability of the produced iron nanoparticles.

The thermal decomposition process using the Schlenk line is effective for producing iron nanoparticles. However, for large-scale production, the integration of Industry 4.0 technologies becomes crucial to oversee the entire process and minimize human intervention. Presently, without adopting Industry 4.0, the existing process is time-consuming, requiring significant manual effort, and operators need to exercise caution when dealing with various chemicals. With the implementation of Industry 4.0, operations become more streamlined, utilizing automated lines and robots to handle materials instead of human workers.

Second Operation - Municipal Wastewater Treatment as a Lights Out Process

This case study focused on the design of a lights-out municipal wastewater treatment facility, analyzing possible improvements, technologies involved, and remaining risks and challenges. The proposed design heavily features eight Industry 4.0 technologies and various sensors to track data needed to run these technologies. Research shows that the addition of Industry 4.0 is plausible in many areas and could largely benefit the process, but there are areas in which it is still impractical or unsafe to fully rely on automation. Emergency, and sometimes routine, maintenance can be unpredictable in wastewater treatment and will likely continue to be done by human hands, but data collection, analysis, material management, and scheduling of maintenance could be improved by proposed "lights out" techniques.

There are over 14,700 wastewater treatment facilities in the United States[13], facilitating treatment of household, industrial, and public space wastewater streams. The process has become increasingly self-sufficient, as only maintenance, sampling, and small adjustments are done by human operators. Our aim was to determine if one could potentially remove human intervention within the bounds of the facility by fully automating wastewater treatment facilities so that they could run as a "lights-out" process, or a process in which, if it were indoors, lights

---

[12] Leena Peltonen, "Practical Guidelines for the Characterization and Quality Control of Pure Drug Nanoparticles and Nano-Cocrystals in the Pharmaceutical Industry," *Advanced Drug Delivery Reviews* 131 (June 1, 2018): 101–15, https://doi.org/10.1016/j.addr.2018.06.009.

[13] Center for Sustainable Systems. University of Michigan. 2023. "U.S. Wastewater Treatment Fact Sheet." Pub. No. CSS04-14. https://css.umich.edu/us-wastewater-treatment-factsheet.

would not be required, as there would be no employees present. This can only be accomplished by using Industry 4.0 technologies.

Yet, we learned that there are areas in which it is still impractical or unsafe to fully rely on automation. Emergency, and sometimes maintenance can be unpredictable in wastewater treatment processes and will likely continue to be done in the foreseeable future by human hands.

A basic understanding of the process was developed through Cole-Parmer's technical resource library[14]. A simplified diagram that summarizes the process is illustrated in figure 3:

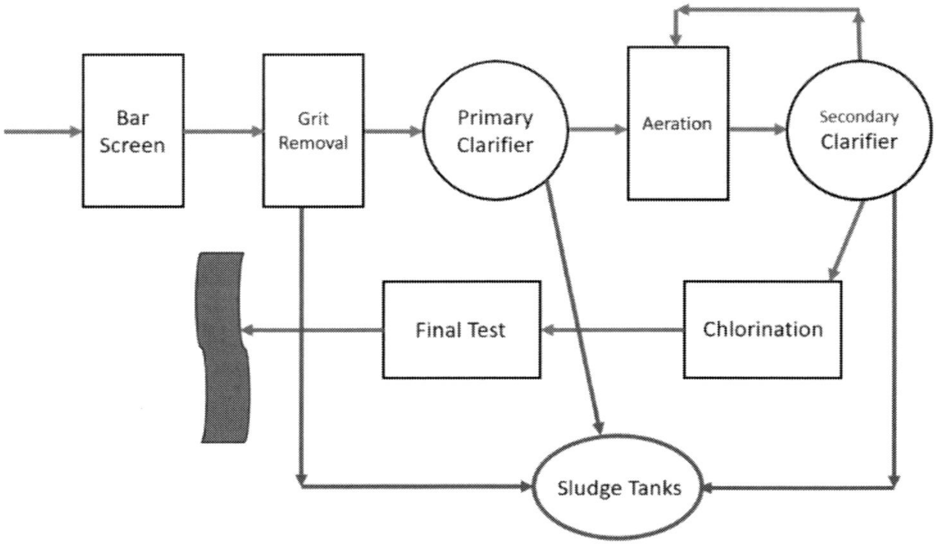

Figure 3. A Simplified Schematic of a Typical Wastewater Treatment Process

Each labeled step is briefly defined below.

- The **bar screen** catches rocks, tree branches, and other large objects that would disrupt the piping system.
- The **grit removal chamber** sifts out sediments, pebbles, and other small particulate through a gravity-based process.
- **Primary clarifiers** remove sludges, oils, and greases from the water, also largely by gravity.
- **Aeration** uses natural bacteria found in wastewater to remove organic matter by feeding it oxygen (air), energizing it and allowing it to break down VOCs.

[14] Cole-Parmer. (2022, September 9). Eight Stages of the Wastewater Process. https://www.coleparmer.com/tech-article/eight-stages-of-wastewater-treatment-process.

- **Secondary clarifiers** serve the same purpose as the primary clarifiers, except the sludge is predominantly biological. There is a recycle line that pumps some of the sludge back into the aeration tank to upkeep the bacteria population.
- **Chlorination**, sometimes called disinfection, involves testing the chemical concentrations and pH of the water and adding chemicals necessary to bring the water to standard.
- **Final test** ensures the water is in compliance with regulations after all cleaning measures are complete before releasing the stream into waterways.
- A **sludge tank**, or series of sludge tanks, collects all sludge and particulate across the process. Sludge can be reused by the plant, sold to another for use, or landfilled.

Other sources that contributed to the "base model" used for this project included the United States Environmental Protection Agency's *How Wastewater Treatment Works... The Basics* from 1998[15]; University of Tennessee Knoxville's Department of Biosystems Engineering Associate Professor John R Buchanan's *Wastewater Basics 101*[16], and NYC Environmental Protection's article "Wastewater Treatment Process," last updated in 2023[17].

Wastewater treatment is already largely self-sufficient, but different plants use design features and methods unique to their process conditions and workforce capabilities. Research was thus done into design options that would be easiest to integrate into a hands-off process.

Through research, a Bar Screen was selected that can withstand a wide range of flows, consistently and automatically rakes trash away from its grates (screen), and requires minimal maintenance, especially for a device with frequent moving parts. The Water Environment Federation's paper "Liquid Stream Fundamentals: Grit Removal"[18] details the advantages and disadvantages of various grit chamber designs. Vortex and mechanical chambers were ruled out to minimize moving part maintenance, favoring aeration or hydraulic chambers. Ecology Center's *Aerated Grit Chamber*[19] article led to a selection of the aerated grit chamber, as it says aerated models remove odors and are favorable for large plants.

---

15 US Environmental Protection Agency. (1998, May). How wastewater treatment works... the Basics – USA EPA. https://www3.epa.gov/npdes/pubs/bastre.pdf.

16 Buchanan, John R. University of Tennessee – Knoxville Agricultural Experiment Station. (2021.) *Wastewater Basics 101*. 19january2021snapshot.epa.gov/sites/static/files/2015-06/documents/epa-mou_wastewater_basics_101.pdf.

17 NYC Environmental Protection. (2023). Wastewater Treatment Process. nyc.gov/site/dep/water/wastewater-treatment-process.page.

18 Water Environment Federation. (2017). Liquid Stream Fundamentals: Grit Removal. integration.wef.org/globalassets/assets-wef/direct-download-library/public/03---resources/wsec-2017-fs-021-mrrdc-lsf-grit-removal_final.pdf.

19 Doty, L. (2023, March 26). Aerated grit chamber – wastewater treatment. Ecology Center. https://www.ecologycenter.us/wastewater-treatment/aerated-grit-chamber.html.

Clarifier research was conducted using Larry W. Moore's presentation *Primary Clarifiers* from the EPA's Operator Webinar series (2023)[20] to select the circular clarifier model, as its circular rake pattern is reliable and the flow of influent comes from the center of the tank, prompting automatic dispersion of untreated water, and Monroe Environmental Corp's "Circular Clarifiers and Thickeners" product database for more information on circular clarifiers.

The US EPA's *The Microbiology of Wastewater Treatment* webinar series from 2020 included a presentation of "Life in the Aeration Tank: Bacteria, Protozoa, and Metazoa" by Jon van Dommelson of the Ohio EPA unit[21]. This presentation provided a better understanding of the bacteria in wastewater and the consequences of not removing it after it is used for biological treatment in aeration tanks. 1H_2O3's article "Bacteria and micro-organisims involved in water treatment"[22] verified that biological treatment can be done without the addition of external bacteria, as long as a fraction of biological sludge is recycled to ensure the bacteria presence in the aeration tank is plentiful. This confirmed that an upstream connection for biological matter was not needed and precautions for its transportation through the plant are unnecessary.

Additional sources were used to understand different steps in the process, select designs, and consider their automation with Industry 4.0. The proposed automation will be discussed in the following section of the report, and other sources used for this purpose are listed in References.

To close the literature search, challenges of the transition to a lights out facility were studied. For example (this will be detailed in **Theory**), this design includes a digital twin of the entire facility to help monitor the plant and detect anomalies. The practicality of a digital twin at this scale was questioned, but an article published in 2022 by ASCE's *Civil Engineering Source* titled "Digital twin being developed for Houston water system"[23] details how Houston Public Works has made a digital twin for three wastewater treatment plants, covering 7500 miles of piping and a 671-square-mile area. While the twin is still in its beta stages, this verifies that a contractor can create a digital twin of this scope. Additionally, Sensorex's "Wastewater Chlorination: Everything You Need To Know" (2022)[24] explains the materials that may be needed for disinfection (upstream), and Amin Almasi's article for *Water Technology* "Sludge Handling in Water & Wastewater Treatment"[25] assisted in research of how sludge is handled (downstream).

---

[20] Moore, L.W. (2023, August 24). Primary Clarifiers. https://www.epa.gov/system/files/documents/2023-09/230824_primary-clarification-for-operators_slides.pdf.
[21] Van Dammelson, J. (2020, November 17). Life in the aeration tank: Bacteria, protozoa, and metazoa. https://epa.gov/sites/default/files/2021-06/documents/microbiologypresentation.pdf.
[22] 1H2O3. (2023). Bacteria and micro-organisms involved in water treatment. 1h2o3.com/en/learn/bacteria-and-microorganisms/.
[23] Cardno, C.A. ASCE. (2022, March). Digital twin being developed for Houston Water System. Civil Engineering Source. https://www.asce.org/publications-and-news/civil-engineering-source/civil-engineering-magazine/issues/magazine-issue/article/2022/03/digital-twin-being-developed-for-houston-water-system.
[24] O'Donnell, D. (2022, June 6). Wastewater chlorination: Everything you need to know. Sensorex Liquid Analysis Technology. https://sensorex.com/wastewater-chlorination/.
[25] Almasi, A. (2021, April 21). Sludge handling in Water & Wastewater treatment. Water Technology. https://www.watertechonline.com/wastewater/article/14185703/sludge-handling-in-water-wastewater-treatment.

Finally, chlorination typically involves sampling and water testing done by humans. Research found that automated water testing robots do exist, as Envirotec reported on a water testing robot by a Dutch company in "New robot for fully automatic water analysis" in 2020[26]. The robot uses spectrometry, UV-VIS, titrations, and gas chromatography to test for pH, coloration, and concentrations of programmed chemicals in water samples. Also, no container material was found that can confidently handle all potential hazards in wastewater samples besides glass. This poses the issue of cracking glass, as robots are not as delicate as gloved human hands. Ultrasonic Technologies advertises that they use ultrasonic vibrations to detect cracks in medical glassware, and a research paper by A. Monastyrski et al, published by IEEE (2008), *Resonance ultrasonic vibrations for in-line crack detection in silicon wafers and solar cells*[27], explains the use of this technology for crack detection in silicon wafers.

The following illustrates the theoretical facility design for a lights out wastewater treatment plant. Figure 4 is a diagram of the new system, followed by table 1, showing which Industry 4.0 technologies are involved in various parts of the proposed system:

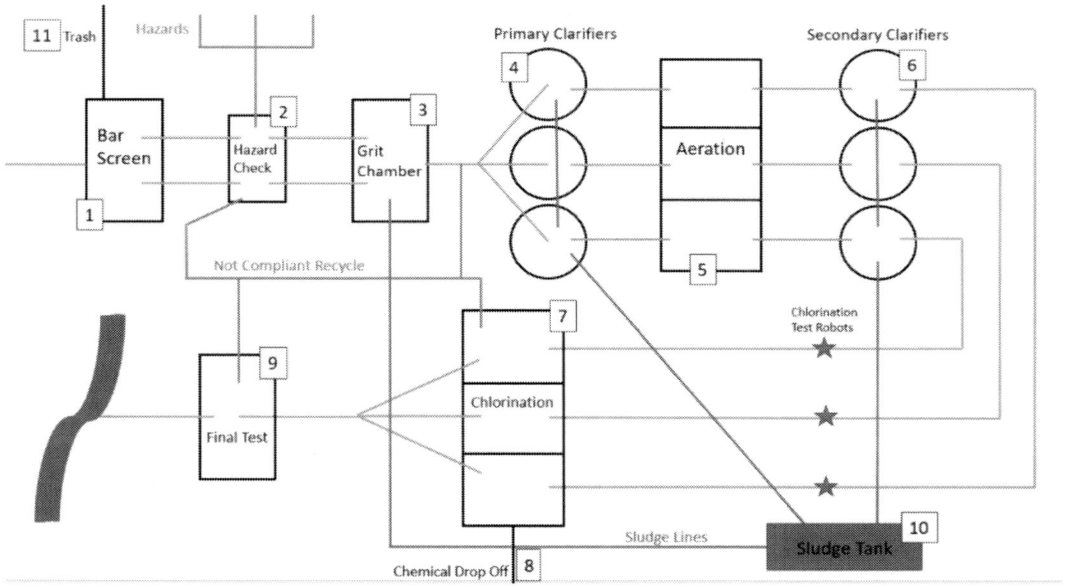

Figure 4. Design of a Lights Out Wastewater Treatment Process

[26] Bickerton, A. (2020, January 17). New Robot for Fully Automatic Water Analysis: Envirotec. Envirotec Magazine. https://envirotecmagazine.com/2020/01/17/new-robot-for-fully-automatic-water-analysis/.

[27] Monastyrski, A., et al. (2008, May). Resonance ultrasonic vibrations for in-line crack detection in silicon wafers and solar cells. IEEE Explore. https://ieeexplore.ieee.org/document/4922742.

Table 1. Industry 4.0 Technologies in Lights Out Process

Technology	Point in Process	Number on Chart
Digital Twin	All	
IoT	Hazard Check, Primary & Secondary Clarifier, Chlorination, Chemical Drop Off, Final Test, Sludge	2, 4, 6, 7, 8, 9, 10
Blockchain	Chlorination, Drop Off	7, 8
Advanced Robotics	Chlorination, Final Test, Trash	7, 9, 11
Machine Learning / Pattern Recognition	Primary & Secondary Clarifier, Final Test	4, 6, 9
AI	Hazard Check, Chlorination, Drop Off, Sludge Tank	2, 7, 8, 10
Touchscreen	Drop Off	8
Big Data	All	

A major difference between the current system and the proposed automated system is the added hazard detection station after large items are physically filtered out. This station will include a HAZMAT device[28] which uses gas chromatography and mass spectrometry to identify such matters like drugs, toxic industrial chemicals, and environmental hazards. Chemicals tested for will be selected based on hazards potentially released by facilities within the zone of the treatment plant. For example, if Company (X) poses the threat of releasing Hazardous Chemical A into their waste stream in an emergency, the device will include Chemical A in its primary testing database. In the event Chemical A is detected by the device, the main line will be closed, and Tank A line opened, diverting the hazardous material to a separate holding tank.

Artificial Intelligence and the Internet of Things will be used to communicate between the device readings and the action taken by the machine, as well as to alert Company (X) that their hazardous material has been detected and needs to be picked up from Tank A.

After the grit chamber, the system splits into three separate lines. Under normal conditions, only the center line is open, but if flow rate is high or flooding is present, additional lines will be opened. Other lines will also be opened as clarifiers require maintenance. Each clarifier will include two level sensors – one for the overall tank, and one for the sludge.

[28] Teledyne FLIR. (2023d). Griffin G510 Person-Portable GC-MS. https://www.flir.com/products/griffin-g510/.

For the sludge, a level detector will be used. This will be a liquid-solid interface level detector that does not require re-calibration with changes in conditions[29]. For the tank level sensors (all clarifiers & tanks), ultrasonic level transmitters will be used.

If the sludge level is too high, this might signify a problem with the clarifier, such as clogging. A temporary solution would be to recycle the wastewater in the clarifier to the pre-clarifier stage and flush the sludge trough, but if it continues being an issue, an external operator will need to be alerted for maintenance. Sludge levels being too low might cause "rat-holing", where clean water is lost through the sludge pump because there is not enough sludge built up in the way. If the sensor reads a low level, a flocculent mixture, such as an aqueous ferric salt, will be pumped into the clarifier. Flocculent encourages sludge particulate to clump together and sink more quickly. This applies to both primary and secondary clarifiers.

Chlorination is the most complex step, as it is the step most often reliant on human judgment. For testing the water, a commercially available robot will be used. The robot readings will be taken at the starred points on the diagram. The data taken by these robots will then be loaded into an IoT data analysis software[30] that can calculate how much of which chemical should be added to adjust concentrations, so the water meets the waterway standards.

Blockchain technology will be implemented here in several ways – each sample reading will add a new block for easy tracking and traceability of skewed readings. Blocks will track the amounts of each chemical delivered and used. Artificial intelligence will inform suppliers when restock of a chemical is needed.

Samples for the robot will be taken in a certified sample jars (500 mL). Although research was done to try and avoid glass, no other material was consistently non-reactive with all possible wastewater components. These jars are certified by the EPA for environmental applications and testing for both organic and inorganic material in water. Because glass is (for now) unavoidable and fragile, a robot will also be employed to gently clean and test for cracks. This robot will use Resonance Ultrasonic Vibration (RUV) technology.

Final testing will use a similar robot to that used for chlorination. If the water is deemed compliant (based on the up-to-date EPA standards, re-reviewed by AI daily), it will be released to the waterway. If it is found to be hazardous, the plant will shut down and it will be reverted to hazard detection. If it is only slightly unclean, it will be recycled to the chlorination step for additional adjustment. Machine learning is employed here, for the robot and AI will eventually "learn" to which point in the process each specific non-compliant reading should be sent.

From upstream suppliers, the plant will need restocks of chemicals – chlorine, acid, base, ferric salt. These will be dropped off at a building reminiscent of a storage unit, where the delivery

[29] Markland. (2023). Automatic Sludge Blanket Level Detector. sludgecontrols.com/our-products/automatic-sludge-level-detector/.
[30] G2. (2023b). Best IOT analytics platforms. https://www.g2.com/categories/iot-analytics-platforms.

person can drop off a chemical in a barrel-like container and input how much of what they are delivering. The system will then accept the barrel and transport the chemical into its own storage tank connected to the chlorination station by pipes and feed pumps.

As for downstream, a contractor will need to drain the sludge tanks and move the sludge to an external handling or disposal location after being alerted by the sludge level sensor. Eventually, sludge recycling could be built into the facility as a source of back-up energy, but this was not a focus of this project. An autonomous dump truck will be used to move larger trash raked from the bar screen to a waste processing facility. A waste processing facility was selected over a landfill because of the possible presence of items not legal for landfills, such as batteries.

The maintenance of the system is the greatest challenge in automating the process, as in many cases, human judgment is required. Equalizer tanks will be employed in case of power outages and forced shutdowns, but operators will often be needed for permanent resolutions. Pattern recognition can be used to signal when routine maintenance is needed. Also, a digital twin will be used to compare ideal conditions to current conditions in case anomalies must be addressed. To upkeep this twin, temperature and pressure, tank levels and volumes, flow rates, and water quality will be needed.

All data will be collected and loaded into an IoT data collector and organizer for industrial-scale applications that include big data analysis capabilities. Figure 5 includes yellow stars to indicate where data readings will be taken for both IoT and the digital twin. Green stars indicate places where the system AI must regularly communicate with external up and downstream points.

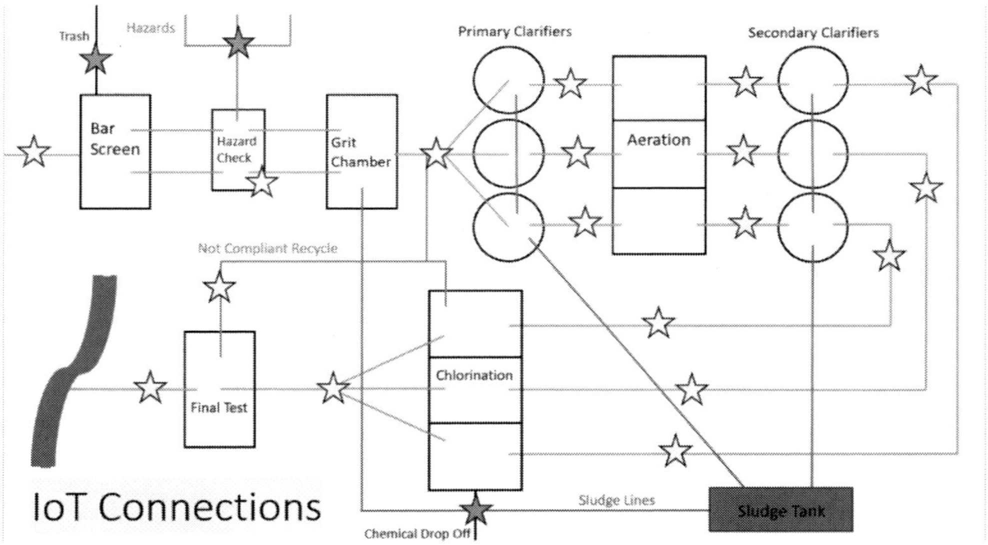

Figure 5. Locations of IoT Connections

While implementations of Industry 4.0 are plausible and could reduce human intervention, certain maintenance requirements, ethical concerns, and the unpredictability of issues at wastewater treatment plants make it complicated to fully automate the process. For instance, equalizer tanks are only a temporary solution of power outages. Releases into waterways or air, tank flooding, pipe clogging, and other serious problems might occur. There are also certain steps in the process, such as the clarifiers, that would be impractical or presently impossible to maintain automatedly.

Logistics aside, environmental and health risks are likewise prevalent in a completely automated system with no human intervention. A malfunction of the hazard detection unit could result in a facility-wide system contamination, and a subsequent waterway contamination if final tests are also skewed. Additionally, bacteria overabundance or stagnation is often determined qualitatively, such as by smell, so a plethora of bacteria could remain undetected and spread waterborne illnesses if released. Ethical debates might also be probed by this proposal, as fully automating wastewater treatment facilities everywhere could leave 10-15 operators and engineers per plant[31] (~15000 plants) unemployed.

However, while there are currently limitations on the extent to which automation can safely and efficiently be used in wastewater treatment, there are benefits to integrating Industry 4.0 into the process. Firstly, it reduces the risk of employee injury or illness. Operators would not regularly be exposed to hazardous components in the water, chemicals used in treatment, or moving mechanical parts. Theoretically, when they are called in for maintenance, AI will have already stopped the moving parts or informed them of the specific hazard they should prepare to encounter. Automation also leads to more consistent judgment. Precise addition of necessary chemicals prevents wasteful overuse of materials. Similarly, records of wastewater concentrations, waterway releases, and system chemical use would be well kept and could be used for future improvements of plants.

At this point, specific numbers are difficult to find for I-4.0 technologies, but clearly automation would be an expensive endeavor. However, considering employee wages and reduced maintenance costs in the long-term, integrating Industry 4.0 could ultimately offer savings.

To conclude, while it may not be practical or safe to truly make a municipal wastewater treatment a lights-out facility, there are many instances where implementation of Industry 4.0 technologies can improve the process in areas of efficiency, cost, and safety. At present, it would be plausible and beneficial to add IoT/AI, blockchain, a digital twin, and robotic chlorination. Upstream, downstream, maintenance, and waterway release will likely remain manual.

All the technologies mentioned are available commercially. This study also suggests the development of a, currently not available, waterproof robot for clarifier cleaning similar in design to the existing robot vacuum cleaners. The capabilities would include vacuuming the floor, wiping

---

[31] DataUSA. (2022). Water & Wastewater Treatment Plant & System Operators. https://datausa.io/profile/soc/water-wastewater-treatment-plant-system-operators.

the impellers, and flushing the sludge trough and influent pipe. Its schedule would be controlled by pattern recognition.

Third Operation - An Injection Molding Process

The objective of here was to investigate the requirements to integrate Industry 4.0 technologies into an Injection Molding (IM) Process and to design a lights-out manufacturing model. The sensors, hardware, and software requirements were investigated to connect the entire system seamlessly together. The upstream and downstream connections were also researched to have a broader scope of the process. Additionally, unsteady processes such as startup, shut down, maintenance and disaster management were considered. Quality control of the process and product inspection were also considered.

Injection Molding (IM) is a process that has mostly been digitized and automated. An IM can adjust its pressure, temperature, velocity, mixing time, cooling time, et cetera[32]. An IM can be set for how long the plastic needs to cool inside of the mold before ejecting. Yet, this is where the automation ends. IM in most industrial settings still requires human interaction in several ways. An IM process cannot detect which settings are correct for each plastic product that it makes. Human intervention is needed to change pressure and temperature settings to ensure that flashing, burning, sticking, and/or clogging does not occur. Additionally, human intervention is required for inspecting the product for defects or anomalies. Not all IM processes use conveyor belts or robots to transport products but are manually loaded and transported by humans. Lastly, safety and security of an IM factory such as fires, malfunctions, break-ins, and routine maintenance are situations demanding human intervention. For lights-out manufacturing to occur Industry 4.0 must address the above challenges.

IoT can be implemented into the IM process by using commercially available devices to connect every part of the process. The IM can transmit data to a Big Data IT Network. This data could be stored on a Cloud Server and Blockchain Technology[33] can keep track of this information as well as custom mold designs and customer data.

Customer data can be kept secure by security software[34] and analyzed by Autonomous Robots to send the product to the correct customer. Digital Twin of an Injection Molding process could be used to predict maintenance, potential anomalies, and allow the process to run smoothly and efficiently. Additive manufacturing of Carbon Steel molds can be used similarly to create products

---

[32] What is Injection Molding and How Does it Work? (2022). YouTube. Retrieved October 4, 2023, from https://youtu.be/sBEzE4PCmqM?si=W5h6VvBoG4j1vW5d.

[33] IBM Linuxone III. IBM. (n.d.). https://www.ibm.com/products/linuxone-iii?mhsrc=ibmsearch_a&mhq=blockchain

[34] Secure your data with HPE Greenlake. HPE. (n.d.). https://www.hpe.com/us/en/solutions/security.html

from metal powder[35]. Collaborative robots (Cobots) can deliver custom molds and replacement parts to the IM machines that require them.

Robots can move the finished product to pallets while cameras inspect the product for defects, rejecting the defected product to be recycled. Pallet Cobots[36] can be used to transport pallets to autonomous delivery trucks and deliver them to the customers. Meanwhile, AI technology can oversee the entire process communicating to each part by IoT[37].

Security cameras can analyze the surroundings and AI can be used to alert authorities in an event of a break-in; a local fire department in an event of a fire; a technician when robots are malfunctioning.

Figure 6 depicts a typical IM process and Table 2 illustrates the corresponding application of Industry 4.0 technologies:

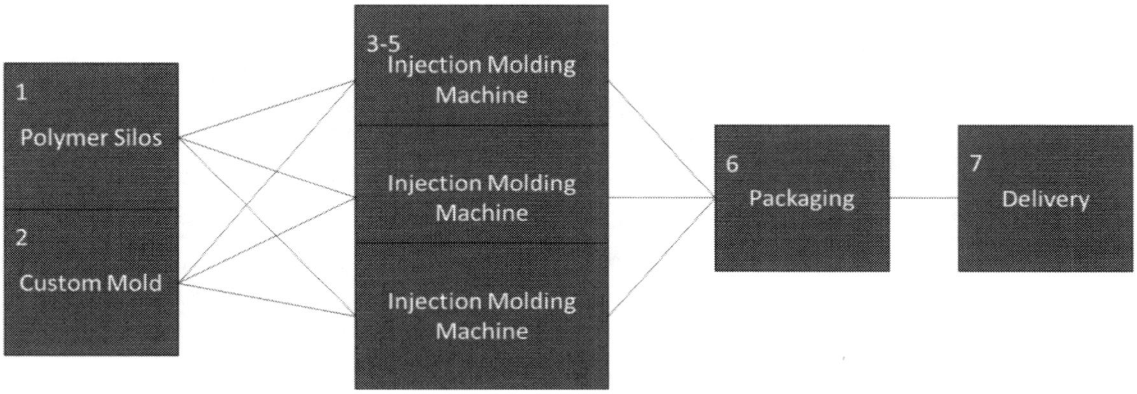

Figure 6. A Typical Injection Molding Process

[35] Additive Manufacturing Services. Sandvik Additive Manufacturing. (n.d.). https://www.additive.sandvik/en/products-services/additive-manufacturing-services/

[36] Fastest, heavy payload cobot. (n.d.). https://www.universal-robots.com/products/ur20-robot/

[37] Okeke, F., Staff, T., Crouse, M., Poremba, S., Clarke, M., & Azhar, A. (2022, October 31). Top 5 industrial IOT use cases and examples. TechRepublic. https://www.techrepublic.com/article/industrial-iot-use-cases/

Table 2. Implementation of Industry 4.0 Technologies

Industry 4.0 Technology	No.
IoT	All
Additive Manufacturing	2
Cloud Storage	All
Artificial Intelligence	1, 2, 6, 7
Autonomous Robots	2, 6, 7
Digital Twin	3, 4, 5
RFID	2, 6, 7
Predicative Maintenance	3, 4, 5
Encryption/Blockchain	All
Cameras	All (s), 5 (i)

In Order to make a plastic product at all, a mold is needed. Using additive manufacturing, a model of the mold can be designed in CAD or similar software[38]. Metal powders can be used to build molds made of carbon steel and cobots can transport the molds into the injection mold (IM) machines. In addition, polymer pellets must be fed into the hopper of the IM machines. These pellets can be transported into the hoppers from silos on site[39]. Part replacement can be determined by digital twinning and predictive maintenance, using IoT to communicate and transport the required replacement part to the correct place. AI can then shut down the system and a cobot can replace the part.

The first part of the injection Molding process is heating the polymer pellets into molten plastic. Next, a large screw mixes and pushes the liquid plastic through[40]. Uniform heating and mixing are essential to prevent burns and uneven filling of the mold. Some companies have already

[38] Wever, P. (2023, May 22). Improve mold machining with equidistant offset, rest finishing and innovative five-axis Toolpath strategies. MoldMaking Technology. https://www.moldmakingtechnology.com/articles/improve-mold-machining-with-equidistant-offset-rest-finishing-and-innovative-five-axis-tool-path-strategies.

[39] Plastic Processing: The Life of a Resin Pellet through Injection Molding and Extrusion. (2020). YouTube. Retrieved November 1, 2023, from https://youtu.be/SgsnZpL0744?si=WkH-tHwnlm5X8GLk.

[40] Injection Molding Animation. (2016). YouTube. Retrieved October 4, 2023, from https://youtu.be/b1U9W4iNDiQ?si=fCfK9P-YOPwhKiZf.

implemented and researched Digital Twins of IM[41]. The digital twin of the system can analyze different scenarios of various temperatures, pressures, flows and mixing speeds to avoid anomalies and find efficient procedures.

After the screw injects the plastic out, it enters the mold at high temperature while cooling water next to the mold removes the heat. Some of the challenges include clogging upon entering the mold, sufficient filling, and uniform cooling of the product. Digital Twinning can also be used to ensure uniform cooling. Using the digital twin, the necessary fill rate of the mold can be determined to ensure correct filling into the mold. To eliminate the clogging, the digital twin can increase the pressure until the clog is removed. To do this however, operations will have to stop, but most clogs typically happen at the startup of operations or at the maintenance time.

The plastic product ejects out of the mold and rotates on a platform. This allows another mold to be filled while the plastic is removed[42]. A nearby cobot grabs the completed product and places it on a pallet. The pallet will have a specific Radio Frequency Identification Tag (RFID) for the product and will communicate to the autonomous pallet robot underneath it to transport it to the correct place[43].

Fire safety is also a crucial factor to consider. One way to consider fire safety is by installing cameras in the factory that use AI to detect fires and signaling the information to digital twins to prevent fires from occurring.

Ultimately the product needs to be shipped to either a customer or another part of the factory that uses the plastic piece to make another product. Before the product can be shipped to the customer, however, the product must be inspected for quality assurance. Using cameras and AI technology, the product can be analyzed for defects, deformities, or other anomalies[44]. The cobot that grabs the product can then be communicated to by IoT to dispose of the defective product to be recycled later. To communicate all required data from the product to the customer, IoT can be used with RFID, which will track the product's location and take it to its desired location. The product can be loaded onto an autonomous truck and taken to the customer.

[41] Injection Molding Machine at Maico Presse S.p.a. with Digital Twin. (2019). YouTube. Retrieved December 1, 2023, from https://youtu.be/MNGI0kpzx6Q?si=aQB7qyxzXdG0xFvt.
https://link.springer.com/chapter/10.1007/978-3-030-49435-3_6

[42] Turnkey system with two interlinked multi-axis robots. YouTube. (2022). YouTube. Retrieved November 1, 2023, from https://www.youtube.com/watch?v=N0y0CmBcMPU.

[43] Industry 4.0: "Smart" luggage tag as a practical example. (2016). YouTube. Retrieved November 1, 2023, from https://youtu.be/57loTeXkxlI?si=QfKqsv-DSgpvN3KF.

[44] Lab equipment. Körber Pharma. (n.d.). https://www.koerberpharma.com/en/solutions/inspection

RFIDs could be implemented onto the IM machines themselves[45]. Robots could communicate with each individual IM machine, determining when maintenance needs to be performed or when malfunctioning occurs. Robots could travel to the machine, scan the RFID and shutdown the system, through a process to perform any maintenance, replace parts. When the robot is finished with maintenance, it can scan the RFID again so that the machine knows it can start back up again.

The implementation of Industry 4.0 for the Injection Molding process has already been implemented in certain companies[46]. Injection Molding may not be the most effective way to produce a variety of custom plastic products, Additive Manufacturing provides the best way to custom tailor plastic products to the specifications provided by the customer.

Results and Conclusions

While the implementation of Industry 4.0 may reduce the need for certain workers, it also presents opportunities for job creation in emerging markets. The shift towards lights-out manufacturing facilities necessitates skilled personnel to oversee operations and monitor performance. This creates a demand for individuals with expertise in handling and optimizing Industry 4.0 technologies. Addressing societal concerns, the ethical implications of job displacement must be balanced with the potential for new job opportunities in the evolving landscape of advanced manufacturing.

In all three cases presented above, key components such as artificial intelligence, the internet of things, and big data play essential roles in facilitating communication and monitoring the overall performance of the processes. While the implementation of these technologies may reduce the need for human operators, they simultaneously introduce new job opportunities being groundbreaking technologies that require knowledge workers to design and maintain them. While certain jobs would be lost, automation could bring new jobs as well. Remote engineers would be needed for system monitoring, and on-call operators would be needed in case of emergency or routine maintenance. Process improvement teams would likely still exist, and subcontractors would still be employed.

As stated earlier, this new course covered various aspects of Industry 4.0. In addition to classroom lectures, reading assignments and tests, student engagement included case studies of selected sample operations in environmental, chemical, and mechanical engineering fields. Students who successfully completed the course acquired the knowledge of how companies might be able to

---

[45] Vulcan RFIDTM label 4" x 2": 4" x 6" – for Zebra and Sato industrial RFID printers. atlasRFIDstore. (n.d.). https://www.atlasrfidstore.com/vulcan-rfid-label-4-x-2-4-x-6-for-zebra-and-sato-industrial-rfid-printers/

[46] Bibow, P., Dalibor, M., Hopmann, C., Mainz, B., Rumpe, B., Schmalzing, D., Schmitz, M., & Wortmann, A. (1970, January 1). Model-driven development of a digital twin for injection molding. SpringerLink.

transform themselves from traditional operations to Industry 4.0 operations and seamlessly incorporate the next generation technologies and agile principles in their modes of operation.

References

[1] Caroline Billings et al., "Magnetic Particle Imaging: Current and Future Applications, Magnetic Nanoparticle Synthesis Methods and Safety Measures," *International Journal of Molecular Sciences* 22, no. 14 (July 17, 2021): 7651, https://doi.org/10.3390/ijms22147651.

[2] Attarad Ali et al., "Synthesis, Characterization, Applications, and Challenges of Iron Oxide Nanoparticles," *Nanotechnology, Science and Applications* Volume 9 (August 1, 2016): 49–67, https://doi.org/10.2147/nsa.s99986.

[3] Cin7, "Inventory Solutions for Product Sellers - CIN7," September 20, 2023, http://www.cin7.com/.

[4] "CYNGN | AI-Powered Autonomous Technology for Industrial Fleets," n.d., http://www.cyngn.com/.

[5] "SIAD Macchine Impianti Air Separation Units | SIAD MI | SIAD Macchine Impianti," n.d., http://www.siadmi.com/air-separation-units.

[6] "Yaskawa Motoman Robotics," Yaskawa Motoman Robotics, n.d., http://www.motoman.com/en-us.

[7] Scott Dalgleish, "Phase IV Engineering | Wireless Sensor Solutions for the Industrial IoT," Phase IV Engineering Inc., December 22, 2022, http://www.phaseivengr.com/.

[8] Chrisnichols, "Automatic Dewar Filling Station," Technifab Products, Inc., June 7, 2023, http://www.technifab.com/filling-station/.

[9] Q-Vac Priming Systems, "Simplex Oil-Lubricated Rotary Vane Priming Systems - Q-VAC Priming Systems," Q-VAC Priming Systems, September 21, 2018, https://vacuumprimingsystems.com/product-center/simplex-priming-systems/simplex-oil-lubricated-rotary-vane-priming-systems/.

[10] "Turn Data into Insights with Oracle Big Data," n.d., http://www.oracle.com/big-data/.

[11] Ultrasonic Technologies, "Glass Crack and Flaw Detection | Ultrasonic Technologies," n.d., https://www.ultrasonictech.com/glass.

[12] Leena Peltonen, "Practical Guidelines for the Characterization and Quality Control of Pure Drug Nanoparticles and Nano-Cocrystals in the Pharmaceutical Industry," *Advanced Drug Delivery Reviews* 131 (June 1, 2018): 101–15, https://doi.org/10.1016/j.addr.2018.06.009.

[13] Center for Sustainable Systems. University of Michigan. 2023. "U.S. Wastewater Treatment Fact Sheet." Pub. No. CSS04-14. https://css.umich.edu/us-wastewater-treatment-factsheet.

[14] Cole-Parmer. (2022, September 9). Eight Stages of the Wastewater Process. https://www.coleparmer.com/tech-article/eight-stages-of-wastewater-treatment-process.

[15] US Environmental Protection Agency. (1998, May). How wastewater treatment works... the Basics – USA EPA. https://www3.epa.gov/npdes/pubs/bastre.pdf.

[16] Buchanan, John R. University of Tennessee – Knoxville Agricultural Experiment Station. (2021.) *Wastewater Basics 101*. 19january2021snapshot.epa.gov/sites/static/files/2015-06/documents/epa-mou_wastewater_basics_101.pdf.

[17] NYC Environmental Protection. (2023). Wastewater Treatment Process. nyc.gov/site/dep/water/wastewater-treatment-process.page.

[18] Water Environment Federation. (2017). Liquid Stream Fundamentals: Grit Removal. integration.wef.org/globalassets/assets-wef/direct-download-library/public/03---resources/wsec-2017-fs-021-mrrdc-lsf-grit-removal_final.pdf.

[19] Doty, L. (2023, March 26). Aerated grit chamber – wastewater treatment. Ecology Center. https://www.ecologycenter.us/wastewater-treatment/aerated-grit-chamber.html.

[20] Moore, L.W. (2023, August 24). Primary Clarifiers. https://www.epa.gov/system/files/documents/2023-09/230824_primary-clarification-for-operators_slides.pdf.

[21] Van Dammelson, J. (2020, November 17). Life in the aeration tank: Bacteria, protozoa, and metazoa. https://epa.gov/sites/default/files/2021-06/documents/microbiologypresentation.pdf.

[22] 1H2O3. (2023). Bacteria and micro-organisms involved in water treatment. 1h2o3.com/en/learn/bacteria-and-microorganisms/.

[23] Cardno, C.A. ASCE. (2022, March). Digital twin being developed for Houston Water System. Civil Engineering Source. https://www.asce.org/publications-and-news/civil-engineering-source/civil-engineering-magazine/issues/magazine-issue/article/2022/03/digital-twin-being-developed-for-houston-water-system.

[24] O'Donnell, D. (2022, June 6). Wastewater chlorination: Everything you need to know. Sensorex Liquid Analysis Technology. https://sensorex.com/wastewater-chlorination/.

[25] Almasi, A. (2021, April 21). Sludge handling in Water & Wastewater treatment. Water Technology. https://www.watertechonline.com/wastewater/article/14185703/sludge-handling-in-water-wastewater-treatment.

[26] Bickerton, A. (2020, January 17). New Robot for Fully Automatic Water Analysis: Envirotec. Envirotec Magazine. https://envirotecmagazine.com/2020/01/17/new-robot-for-fully-automatic-water-analysis/.

[27] Monastyrski, A., et al. (2008, May). Resonance ultrasonic vibrations for in-line crack detection in silicon wafers and solar cells. IEEE Explore. https://ieeexplore.ieee.org/document/4922742.

[28] Teledyne FLIR. (2023d). Griffin G510 Person-Portable GC-MS. https://www.flir.com/products/griffin-g510/.

[29] Markland. (2023). Automatic Sludge Blanket Level Detector. sludgecontrols.com/our-products/automatic-sludge-level-detector/.

[30] G2. (2023b). Best IOT analytics platforms. https://www.g2.com/categories/iot-analytics-platforms.

[31] DataUSA. (2022). Water & Wastewater Treatment Plant & System Operators. https://datausa.io/profile/soc/water-wastewater-treatment-plant-system-operators.

[32] What is Injection Molding and How Does it Work? (2022). YouTube. Retrieved October 4, 2023, from https://youtu.be/sBEzE4PCmqM?si=W5h6VvBoG4j1vW5d.

[33] IBM Linuxone III. IBM. (n.d.). https://www.ibm.com/products/linuxone-iii?mhsrc=ibmsearch_a&mhq=blockchain

[34] Secure your data with HPE Greenlake. HPE. (n.d.). https://www.hpe.com/us/en/solutions/security.html

[35] Additive Manufacturing Services. Sandvik Additive Manufacturing. (n.d.). https://www.additive.sandvik/en/products-services/additive-manufacturing-services/

[36] Fastest, heavy payload cobot. (n.d.). https://www.universal-robots.com/products/ur20-robot/

[37] Okeke, F., Staff, T., Crouse, M., Poremba, S., Clarke, M., & Azhar, A. (2022, October 31). Top 5 industrial IOT use cases and examples. TechRepublic. https://www.techrepublic.com/article/industrial-iot-use-cases/

[38] Wever, P. (2023, May 22). Improve mold machining with equidistant offset, rest finishing and innovative five-axis Toolpath strategies. MoldMaking Technology. https://www.moldmakingtechnology.com/articles/improve-mold-machining-with-equidistant-offset-rest-finishing-and-innovative-five-axis-tool-path-strategies.

[39] Plastic Processing: The Life of a Resin Pellet through Injection Molding and Extrusion. (2020). YouTube. Retrieved November 1, 2023, from https://youtu.be/SgsnZpL0744?si=WkH-tHwnlm5X8GLk.

[40] Injection Molding Animation. (2016). YouTube. Retrieved October 4, 2023, from https://youtu.be/b1U9W4iNDiQ?si=fCfK9P-YOPwhKiZf.

[41] Injection Molding Machine at Maico Presse S.p.a. with Digital Twin. (2019). YouTube. Retrieved December 1, 2023, from https://youtu.be/MNGI0kpzx6Q?si=aQB7qyxzXdG0xFvt.
https://link.springer.com/chapter/10.1007/978-3-030-49435-3_6

[42] Turnkey system with two interlinked multi-axis robots. YouTube. (2022). YouTube. Retrieved November 1, 2023, from https://www.youtube.com/watch?v=N0y0CmBcMPU.

[43] Industry 4.0: "Smart" luggage tag as a practical example. (2016). YouTube. Retrieved November 1, 2023, from https://youtu.be/57loTeXkxlI?si=QfKqsv-DSgpvN3KF.

[44] Lab equipment. Körber Pharma. (n.d.). https://www.koerberpharma.com/en/solutions/inspection

[45] Vulcan RFIDTM label 4" x 2": 4" x 6" – for Zebra and Sato industrial RFID printers. atlasRFIDstore. (n.d.). https://www.atlasrfidstore.com/vulcan-rfid-label-4-x-2-4-x-6-for-zebra-and-sato-industrial-rfid-printers/

[46] Bibow, P., Dalibor, M., Hopmann, C., Mainz, B., Rumpe, B., Schmalzing, D., Schmitz, M., & Wortmann, A. (1970, January 1). Model-driven development of a digital twin for injection molding. SpringerLink.

69a Online Sensing of Real-Time Paramagnetic Properties for Flow Assurance Management

John Lovell, Omar Kulbrandstad

As anticipated by Lord Kelvin, an inability to measure a certain component creates challenges and, perhaps opportunities, for optimization. This has proven true in aspects of the chemical management of flow assurance in the oilfield, wherein asphaltenic particles precipitate and cause significant fouling and obstruction to flow. The oilfield industry has developed measurement techniques on aspects of asphaltenes in solution, but has not previously been able to make online measurements pertinent to the asphaltene itself.

This changed with the discovery that the paramagnetism of asphaltenic molecules did not depend on whether those molecules were in solution, flocculated in "near-solution", or precipitated into larger constructs. As a result, a real-time measurement of paramagnetism can be used to infer conveyance of asphaltene upstream of that measurement. In particular, a surface measurement of asphaltene at the wellhead can be used to optimize a chemical treatment program in the subsurface.

An optimal solution to measuring real-time paramagnetism has been developed based on the recent concepts of big-data analysis. Instead of measuring the quantity of a certain component once per year, instead measures quantum chemical properties every minute as the fluid flows past the sensor. Such a measurement is inherently noisy but AI techniques can be used to extract pertinent features, such as the concentration of a certain paramagnetic species or hyperfine spacing within the species. One key point is that the fluid remains at wellhead temperature and pressure and there is no opportunity for oxygen or other contaminants to degrade the signal.

Such a combination of sensor, real-time digitization and IoT (Internet-of-Things) was developed for flow assurance applications in 2017 and is now generating data in Abu Dhabi, Gulf-of-Mexico and Central America. The device works by passing well fluid through a chamber that can be excited with ~4GHz

RF

field and a swept DC magnetic field of up to ~1500 Gauss. Unpaired electrons in an orbital will resonate

at a certain ratio of Gauss/GHz where that ratio is representative of the interaction between the electron

and nearby nuclei. In the case of asphaltene, the dominant signal comes from electrons on conjugated

double bonds (π bonds) found in the aromatic rings and other structural elements. Additional

paramagnetic features can also appear from vanadium, sulfur and iron within the molecule.

The overall spectrum is transmitted by the IoT software to the cloud where machine learning based

techniques are used to extract the π-bond signature. The peak-to-peak value of the signature has been

shown in laboratory testing to directly comparable with the traditional gravimetric methods (SARA, etc.)

that have traditionally been used in the industry. One surprise is that the measurement turns out to be

quite dynamic. For example the asphaltene might alternatively deposit inside a pipe and then slough off,

which gives a sinusoidal variation in the spectral amplitude. Initial results indicate that some wells can

give a characteristic dynamic pattern in the few days before the wellhead completely seizes up due to

deposition – a result that gives significantly earlier warning than can be seen from pressure gauges.

Operators are gaining confidence in the measurement and beginning to incorporate the digital data into

control loops to optimize their chemical management process. It also seems reasonable to infer that realtime

paramagnetic measurements should also have applicability to fouling and deposition within surface

refineries and chemical plants. The combination of quantum chemical paramagnetic sensing, combined

with real-time IOT, cloud storage and machine learning led to the sensor winning the iChemE Global

Research and ADIPEC Best Innovation awards in 2021

69b Machine Learning-Aided Graphene Chemisensors for Food Safety Monitoring

Sarbashis Das, Andrew Pannone, Harikrishnan Ravichandran, Saptarshi Das

Ion-sensitive field-effect transistors (ISFETs) have emerged as indispensable tools in chemosensing applications due to their remarkable precision and sensitivity. ISFETs operate by converting changes in the concentration/composition of chemical solutions into electrical signals, making them ideal for environmental monitoring, healthcare diagnostics, and industrial process control. Recent advancements in ISFET technology, including functionalized multiplexed arrays and advanced data analytics, have improved their performance. In this study, we illustrate the advantages of incorporating machine learning (ML) algorithms to construct predictive models using the extensive datasets generated by ISFET sensors. This integration enables us to extract intricate patterns that surpass what can be derived solely from human expertise. Furthermore, it addresses common challenges, including cycle-to-cycle, sensor-to-sensor, and chip-to-chip variations, thereby alleviating the manufacturing complexities associated with achieving flawless sensors. As a result, this paves the way for the broader adoption of ISFETs in commercial applications. Specifically, we utilize non-functionalized graphene-based chemitransistor arrays to train artificial neural networks (ANNs) with a remarkable ability to discern instances of food fraud, food spoilage, and food safety concerns. We anticipate that the fusion of compact, energy-efficient, and reusable graphene-based ISFET technology with robust machine learning algorithms holds the potential to revolutionize the detection of subtle chemical environmental changes, offering swift, data-driven insights applicable across a wide spectrum of applications.

69c Ensuring Preventive and Mitigative Measures in F BF Confined Space through Online Measurements for Toxic and Flammable Gases

Abhineet Raj, Naveen Kumar

Background :

Steel industries are hazardous in nature due to presence of toxic and flammable gases which are generated during processes such as iron making in blast furnace, coke making using coal and steel making in basic oxygen convertors. The toxic gas in steel industries is mostly carbon monoxide gas which is present in by product gas of blast furnace process and steel making process. There are several jobs being executed during shutdown and many to those need to be performed in confined space. The personal safety of workmen is of utmost importance and using a personal gas detection system for identifying location specific hazards are not sufficient. Some of the confined space jobs are need to carried out without isolating the line from both side and this poses a greater risk of gassing and asphyxiation. The toxicity hazard is not the only hazard present , some of cases the gas composition is flammable as well as toxic such as by product gas generated during coke making process called as coke oven gas. It consist of mainly hydrogen and hydrocarbon along with carbon monoxide. The presence of a complete gas free environment along with a workable oxygen level is must for working in such condition.

Implementation : To ensure a workable environment in confined space it becomes mandatory to not only do the measurement of gases and oxygen concentration but also preventive and mitigative safeguards to be built based on alert generated by these measurement. To ensure this a continuous online gas measurement system is incorporated which is a sensor based analyser and may help in detecting presence of gases as per pre-set value and also capable of generating alerts in the form of hooter and flasher light. These area monitoring devices also detects the level of Lower explosive limit and triggers an alert to stop the hot work going inside the confined space.

Advantages:

1. **Innovative area monitoring**

Innovative area monitoring devices for a wide range of applications. An outstanding combination for increased safety – in addition to the personal air monitors carried on the body, these area monitors are positioned where gas hazards are expected.

2. Clear warning

Even from a distance, the illuminated green LED ring indicates that the air is free of toxic and combustible gases. Upon detection of gas hazards, the LED color changes from green to red, thus providing a clear visual warning that gas is present. Additionally, a loud and highly audible evacuation alarm is emitted.

3. Wireless fence line

Up to 25 devices can be automatically interconnected to form a wireless fence line. The interconnection of the area monitoring devices allows for rapid establishment of safety networks for larger areas.

4. This gas detection instruments can be used for the measurement of up to six gases and extends the portable gas detection technology to a comprehensive system with many applications.

43a Panel discussion: NLP and LLM in the Chemical Industry

Michael Dessauer, Mark Pietryka, Brent Railey, Greg Parkison

Generative AI tools have proven to hold remarkable capabilities that aims to augment and possibly disrupt knowledge worker roles in many industries, with process manufacturing being no exception. In this discussion, panelists from industry and vendor groups will share real-world case studies and the challenges of integrating Generative AI into process manufacturing workflows. The panel discussion session explores how Generative AI can revolutionize process design, optimization, and operations that delivers value for companies. Please join us to learn more on how process industry is harnessing the power of Generative AI for predictive modeling, process optimization, workflow automation, and intelligent decision-making

192a Valuable Generative AI Applications for the Chemical Industry

Sina Pakazad, Timothy Holtan

Attendees of **Valuable GenAI Applications for the Chemical Industry** will learn how C3 AI developed and applied Generative AI applications to drive Industry transformation. C3 AI will first highlight a case study describing Generative AI applications to analyze time series data and increase revenue. Second, C3 AI will share a case study describing how a steel company improved plant safety outcomes with an innovative Generative AI application.

192b Accelerated Process and Product Development with Small Data and Machine-Learning

Daniela Blanco, Ethan Tenison, César Urbina-Blanco

Data-driven machine-learning (ML) offers the possibility to dramatically reduce the resources needed for reaction, formulation, and process optimization through predictions of behavioral trends, variable effects, and optimal conditions. However, most ML algorithms require large datasets (i.e. 100's-1000's of points) to provide accurate predictions, and experimental data collection is expensive in human and material resources. The challenge of building large datasets can be averted with Bayesian optimization (BO) frameworks. These guide experimental campaigns and carefully build training datasets that ultimately reduce the number of experiments required to achieve reliable understanding of variable effects and reaction behavior. However, their implementation is not straightforward in many cases.

Developing an enhanced BO approach, we elucidate a reaction-agnostic pathway for the design and implementation of intelligent experimental campaigns using material formulations, as well as catalytic and electrochemical transformations as case studies. Working at the intersection of ML and chemistry, we created SuntheticsML, an online, user-friendly, ML platform available to researchers that does not require coding or ML expertise. It leverages small datasets (starting with only 5 data points) to enable the development of new chemical processes, materials, and formulations with **unprecedented efficiencies** and **up to 32x less time and experiments.**

The insights from this work innovate on the future of reaction engineering, material design, and process optimization while lowering the barrier to the implementation of new chemical technologies. SuntheticsML accelerates lengthy and complex traditional optimization campaigns, harnessing the power of ML and small data to fast-track innovation, sustainability, and digitalization in the chemical industry, further reducing development waste, emissions, and resource consumption by **up to 97%.**

192c Industrial AI: Industrial Applications of Machine Learning and Artificial Intelligence

David Vickery, Katherine Cobb

The Process Industries are awash in "lakes" of data – although "oceans" may be a more apt description: data from operations, data from labs, data from supply chains, and so on. The challenge is how to derive the greatest benefit from that data.

Aspen Technology is helping customers do just that – derive benefits from all this data by embedding Industrial AI in its applications.=

By embedding AI in our solutions; we provide fit for purpose, domain-specific application of AI in an industrial context. This enables data scientists to collaborate with domain experts and create solutions that maximize the benefits derived from that data. We do this by taking the pervasive data from across your enterprise and value chain, and applying a combination of

- **Intelligence**, in the form of AI/ML capabilities and other advanced analytics, and
- **Domain Expertise**, the first principles engineering (physics and chemistry) at the heart of these highly complex assets processes

to deliver actionable insights that power measurable business outcomes.

In this presentation, we will touch on how these AI/ML techniques are embedded in our various solutions spaces. We will then highlight, and demonstrate, how Aspen Hybrid™ Models can be used to create models of processes for which simulation models are not available or to improve the ability of first principle models to match real plant/equipment operations.

90a Improving LLM Understanding for Process Manufacturing: Successful Approaches and Lessons Learned

Greg Parkison, Michael Dessauer

As process manufacturing industries begin to develop domain-aware LLM applications, users and developers may struggle with LLM understanding of technical and domain-specific language. The misunderstanding of uniquely identified entities and procedures will pose challenges to accurately retrieve relevant documents and perform general document understanding tasks. The Westlake Digital team has developed custom workflows and functions to augment user prompts to assist LLM language understanding and document retrieval tasks. In this talk, we will share our approach and key learnings on how to improve LLM application performance for operations.

90b Transforming Chemical Manufacturing Decision Making with Retrieval Augmented Generation (RAG): Grounding AI Responses with Domain Specific Data

Justin Kauhl

Facing the contemporary challenges in the chemical manufacturing sector, the integration of advanced computational technologies becomes more critical than ever. This presentation aims to discuss the novel technique of Resource Augmented Generation (RAG), a state-of-the-art that transforms how generative AI can formulate responses without the necessity for expensive retraining or fine-tuning of Large Language Models (LLMs).

RAG is an emergent solution that bridges the gap between AI capacity and specialized knowledge, by grounding AI-generated responses on domain specific data sources. Our method integrates RAG into strategic decision-making workflows by aligning it with our existing databases, thus effectively leveraging AI's potential to produce contextual responses on otherwise unfamiliar subjects.

We intend to share real examples of how RAG is used in our chemical manufacturing business' strategic planning. Even at the early stages integrating RAG boosted GenAI into everyday decision-making processes, has improved our overall efficiency and allowed us to make more informed decisions. It's our hope that these cases will showcase how leveraging GenAI and RAG can have a tangible impact on chemical manufacturing strategies.

90c How Natural Language Processing Boosts Knowledge Sharing and Productivity in the Chemical Manufacturing Industry

Felix Monschein

Ai, and specifically Natural Language Processing (NLP) is changing the way chemical manufacturers are operating--from shop floor to its corporate headquarters. Industry 4.0 has significantly increased the amount of data that is available to operators but how do they discover relevant information? Smart searching developments are helping companies like Bayer to discover critical information quickly and efficiently by using a search feature that is designed for the operations and using the latest AI-based language models. In this presentation, we will show how Bayer deployed a smart search engine that significantly increased their productivity.

The development of an effective and efficient search functionality based on AI looked to solve two challenges cited by Bayer:

1. Inefficient access to documented historical knowledge captured by Shiftconnector; and
2. Incomplete entry items where often descriptions of solutions to problems (in the production process) were missing or incompletely described. The development of AI's Smart Search functionality solved both problems. Teams have improved plant performance by quickly finding solutions based on the history of documented tribal knowledge from plant operations teams. By retrieving the most relevant information, it accelerates operations and brings capabilities to help teams recover from disruptions by immediately identifying appropriate fixes. Also, captured knowledge can be managed to support workforce changes.

The development of the Smart Search feature allows the search engine to

feature to understand the context of the search versus a standard keyword search that would need very specific terms to retrieve relevant information. With Machine Learning, the smart search retrieves relevant information with a simple, everyday phrase like "why is this product brown instead of gray?" Search engines in today's market like Google are smart but do not think like someone in the process industry as they lack industry-specific or even plant-specific terminology.

Smart Search is superior in this aspect as it was developed with a lexicon of the operators in the process industry. Working closely with Bayer operators that had in-depth knowledge of their processes and solutions, the smart search engine was developed to understand not only the specific use cases for a search engine but also the specific terminology that is used every day on the plant floor. Consequently, Smart Search covers plant-specific terminology, such as searching for technical labels which are only used at a certain production line. For example, if in a plant the label "CP001" is used for a centrifugal pump at a reactor inlet, Smart Search understands this and will retrieve relevant information linked to "CP001" when an operator is searching for "centrifugal inlet pump".

Analysis of the deployment shows:

- Smart Search is well accepted and used several times daily.
- The time users spend searching for information decreased significantly. Instead of searching for several minutes, users get results within seconds.
- User statistics show that users access all historical knowledge (> 8 years) captured. Prior to the Smart Search introduction, only records from the last month were accessed.

Statistics and questionnaires show that now the historical know-how captured (solutions to problems in the production process) is effectively used by all users.

165a Immersive Virtual Reality Applications in Safety Training

Abdulrahman Alamer, Asma Alromaih, Abdulmalik Alghanmi, Lama Aldosary

The oil and gas industry often requires people to work in hazardous environments and to operate complicated equipment which often limits the type of training that is carried out on site. The conventional safety training imposes several challenges both in terms of cost and risk. More often than not, it is unsafe or even dangerous to expose trainees to real working environments for the purposes of training. This fact has motivated safety professionals to explore further innovative ways to remedy the problem. Immersive Virtual Reality (IVR) has been proposed as a technological breakthrough that holds the power to facilitate learning, which exposes personnel to hazardous situations in a safe, highly visual and interactive manner. IVR has been proven to enhance understanding, improve knowledge retention, and provides realistic scenarios tailored towards the audience.

165c A View into Industrial Metaverse

Iiro Esko, Andrea Sassetti

Join this session to gain insight into what Industrial Metaverse looks like today. We will look at two industrial facility visualizations, the baseline and definition of Industrial Metaverse, executable digital twins, user interfaces and use cases, challenges and areas of development.

Components that are considered in this abstract:

Visualization and insights from a realistic environment - 3D visualization on screen or on extended reality devices creating the immersion of being pulled into an environment. Getting feedback from real-world assets & their digital twins regarding the right info, right time, and right experts. Enabling multiple contributors, multiple disciplines working on the single source of truth in a next level experience.

Meeting in real-time with immediate collaborative review and changes - Featuring 3D representation of all components of a plant, embedded engineering directly from the 3D application, AI analysis and root cause guidance, supply chain delivered component models and integration, virtual commissioning collaborations, relevant real-time simulation as needed, possibility to zoom in and out of different abstraction levels individually and at once for all levels.

Continuous evaluation, simulation, and prediction - Talking about model order reduction of most accurate models to lowered computational complexity while conserving underlying physics to allow real-time execution on edge devices with lower hardware footprint. Other angles include physics-aware surrogate-based approaches and GPU acceleration saving time without sacrificing accuracy for dedicated problem classes.

Monitor, analyze, and manage assets - XiL approaches like Hardware-in-Loop (HiL) and Software-in-Loop (SiL) for computer aided manufacturing enable time savings for energy, feedstock, maintenance, optimization, and quality related operations in a safe and efficient way.

This session is aimed at technologists interested in photorealistic real-time environments applied through scalable computing in real-time enabling 3D collaboration across a comprehensive digital twin representing mechanical, electrical, and software domains of the virtual and physical realities.

179b From Theory to Impact: Unlocking the Power of Bayesian Optimization on Real-World Science and Engineering Systems

Joel Paulson

Bayesian optimization (BO) is a powerful tool for optimizing non-convex black-box functions that are expensive and/or time-consuming to evaluate and may be subject to stochastic noise in their evaluations. Many important problems can be formulated in this manner, such as optimizing outcomes of high-fidelity computer simulations, hyperparameter tuning in machine learning algorithms, A/B testing for website design, policy-based reinforcement learning, process flowsheet synthesis, and material and drug discovery. In this presentation, three key concepts are introduced, which we argue are critical for enabling or improving the performance of BO on real-world science and engineering systems. Specifically, one must: (1) leverage prior (physics-based) knowledge to perform highly efficient, targeted exploration of the solution space; (2) explicitly incorporate safety constraints during interaction with physical systems to avoid unsafe/undesirable outcomes; and (3) account for external uncertainty sources to ensure the best-found solution is robust/flexible in practice. We introduce a unified framework for adapting BO to handle these considerations and illustrate how this framework can be deployed in practice on a series of examples ranging from the design of safe cold plasma jet devices to the discovery of high-performance sustainable energy storage materials. We also offer perspectives on key challenges and future opportunities in the realm of applied BO.

AUTHOR INDEX